# Advances in Isotope Methods for the Analysis of Trace Elements in Man

# CRC SERIES IN MODERN NUTRITION

## Edited by Ira Wolinsky and James F. Hickson, Jr.

**Published Titles**

*Manganese in Health and Disease*, Dorothy J. Klimis-Tavantzis
*Nutrition and AIDS: Effects and Treatments*, Ronald R. Watson
*Nutrition Care for HIV-Positive Persons: A Manual for Individuals and Their Caregivers*, Saroj M. Bahl and James F. Hickson, Jr.
*Calcium and Phosphorus in Health and Disease*, John J.B. Anderson and Sanford C. Garner

## Edited by Ira Wolinsky

**Published Titles**

*Practical Handbook of Nutrition in Clinical Practice*, Donald F. Kirby and Stanley J. Dudrick
*Handbook of Dairy Foods and Nutrition*, Gregory D. Miller, Judith K. Jarvis, and Lois D. McBean
*Advanced Nutrition: Macronutrients*, Carolyn D. Berdanier
*Childhood Nutrition*, Fima Lifschitz
*Nutrition and Health: Topics and Controversies*, Felix Bronner
*Nutrition and Cancer Prevention*, Ronald R. Watson and Siraj I. Mufti
*Nutritional Concerns of Women*, Ira Wolinsky and Dorothy J. Klimis-Tavantzis
*Nutrients and Gene Expression: Clinical Aspects*, Carolyn D. Berdanier
*Antioxidants and Disease Prevention*, Harinda S. Garewal
*Advanced Nutrition: Micronutrients*, Carolyn D. Berdanier
*Nutrition and Women's Cancers*, Barbara Pence and Dale M. Dunn
*Nutrients and Foods in AIDS*, Ronald R. Watson
*Nutrition: Chemistry and Biology, Second Edition*, Julian E. Spallholz, L. Mallory Boylan, and Judy A. Driskell
*Melatonin in the Promotion of Health*, Ronald R. Watson
*Nutritional and Environmental Influences on the Eye*, Allen Taylor
*Laboratory Tests for the Assessment of Nutritional Status, Second Edition*, H.E. Sauberlich
*Advanced Human Nutrition*, Robert E.C. Wildman and Denis M. Medeiros
*Handbook of Dairy Foods and Nutrition, Second Edition*, Gregory D. Miller, Judith K. Jarvis, and Lois D. McBean
*Nutrition in Space Flight and Weightlessness Models*, Helen W. Lane and Dale A. Schoeller

*Eating Disorders in Women and Children: Prevention, Stress Management, and Treatment,* Jacalyn J. Robert-McComb
*Childhood Obesity: Prevention and Treatment,* Jana Pařízková and Andrew Hills
*Alcohol and Coffee Use in the Aging,* Ronald R. Watson
*Handbook of Nutrition and the Aged, Third Edition,* Ronald R. Watson
*Vegetables, Fruits, and Herbs in Health Promotion,* Ronald R. Watson
*Nutrition and AIDS, Second Edition,* Ronald R. Watson
*Advances in Isotope Methods for the Analysis of Trace Elements in Man,* Nicola Lowe and Malcolm Jackson
*Nutritional Anemias,* Usha Ramakrishnan
*Handbook of Nutraceuticals and Functional Foods,* Robert E. C. Wildman

**Forthcoming Titles**

*Nutrition for Vegetarians,* Joan Sabate
*Tryptophan: Biochemicals and Health Implications,* Herschel Sidransky
*The Mediterranean Diet,* Antonia L. Matalas, Antonios Zampelas, Vasilis Stavrinos, and Ira Wolinsky
*Handbook of Nutraceuticals and Nutritional Supplements and Pharmaceuticals,* Robert E. C. Wildman
*Insulin and Oligofructose: Functional Food Ingredients,* Marcel B. Roberfroid
*Micronutrients and HIV Infection,* Henrik Friis
*Nutrition Gene Interactions in Health and Disease,* Niama M. Moussa and Carolyn D. Berdanier

# Advances in Isotope Methods for the Analysis of Trace Elements in Man

*Edited by*
*Nicola Lowe, Ph.D. and*
*Malcolm Jackson, Ph.D.*

CRC Press
Boca Raton London New York Washington, D.C.

**Library of Congress Cataloging-in-Publication Data**

Jackson, Malcolm J.
Advances in isotope methods for the analysis of trace elements in man / by Malcolm J. Jackson, Nicola M. Lowe.
p. cm. — (CRC series in modern nutrition)
Includes bibliographical references and index.
ISBN 0-8493-8730-2 (alk. paper)
1. Trace elements—Analysis. 2. Trace elements in human nutrition. 3. Trace elements—Isotopes. I. Lowe, Nicola M. II. Title. III. Modern nutrition (Boca Raton, Fla.)

QP534.J33 2000
613.2'8—dc21 00-058562
CIP



Direct all inquiries to CRC Press LLC, 2000 N.W. Corporate Blvd., Boca Raton, Florida 33431.


International Standard Book Number 0-8493-8730-2
Library of Congress Card Number 00-058562
Printed in the United States of America 1 2 3 4 5 6 7 8 9 0
Printed on acid-free paper

## SERIES PREFACE FOR MODERN NUTRITION

The CRC Series in Modern Nutrition is dedicated to providing the widest possible coverage of topics in nutrition. Nutrition is an interdisciplinary, interprofessional field par excellence. It is noted by its broad range and diversity. We trust the titles and authorship in this series will reflect that range and diversity.

Published for a scholarly audience, the volumes in the CRC Series in Modern Nutrition are designed to explain, review, and explore present knowledge and recent trends, developments, and advances in nutrition. As such, they will also appeal to the educated general reader. The format for the series will vary with the needs of the author and the topic, including, but not limited to, edited volumes, monographs, handbooks, and texts.

Contributors from any bona fide area of nutrition, including the controversial, are welcome.

We welcome this important and timely contribution to this series. This book will be useful to a broad spectrum of nutritionists and life scientists of all walks.

**Ira Wolinsky, Ph.D.**
*University of Houston*
*Series Editor*

# Preface

Developments in isotope methods for studying trace elements have reached the stage where we can now use isotopes to answer many questions about status, absorption, turnover, etc., that are inaccessible by other techniques. Nevertheless, the use of isotopes and, particularly, stable isotopes remains an approach followed by only a minority of investigators in this exciting area of human nutrition. Our aim in publishing the group of authoritative reviews in this book is to bring to a wider audience the large potential of these approaches, and to provide definitive information on trace element status and metabolism.

The chapters are a state-of-the-art collection from leading experts in this field, and from investigators in Europe and the United States, reflecting the relatively high cost of establishing isotope analysis facilities. Cost has undoubtedly been one of the major factors limiting widespread use of stable isotopes, particularly purified isotopes and specialized mass spectrometry facilities. Nevertheless, one of the aims of this book is to demonstrate that these costs are justified. The field has developed sufficiently so that validated experimental approaches are available and applicable to studies in a wide variety of subjects, such as in underdeveloped countries or to specific patient groups. The benefit that can be accrued from such studies is substantial.

It is apparent from the chapters presented here that investigators in this field are excited by the potential of isotope techniques to inform our research in trace-element nutrition and metabolism. We hope that readers will be stimulated to pursue these approaches in their research. Finally, we would like to thank our chapter contributors for their help and patience during the development of this project.

**Nicola M. Lowe**
**Malcolm J. Jackson**
**Liverpool, U.K.**

# Editors

**Nicola M. Lowe, Ph.D.,** is a Senior Lecturer in Nutrition at the University of Central Lancashire, U.K. She holds a joint honours degree in Biochemistry and Nutrition from the University College of North Wales, Bangor, and a Ph.D. degree from the University of Liverpool, Department of Medicine. During her doctoral research, Dr. Lowe developed stable isotope techniques, coupled with mathematical modelling to study zinc metabolism and kinetics in humans. After completing her Ph.D., Dr. Lowe joined the team lead by Professor Janet King in the Department of Nutritional Sciences at Berkeley, California. She spent 4 years at Berkeley as a postdoctoral researcher, where she continued her work in the field of stable isotope studies of zinc metabolism. Her current research activities include the use of stable isotope techniques to study zinc and copper kinetics in patients with osteoporosis, and selenium status in a U.K. population. Dr. Lowe has published several papers on trace element metabolism in the *American Journal of Clinical Nutrition* and the *British Journal of Nutrition,* and is a member of the Nutrition Society.

**Malcolm J. Jackson, Ph.D.,** is Professor of Cellular Pathophysiology and Head of the Department of Medicine at the University of Liverpool, U.K. He holds a B.Sc. honours degree from the University of Surrey, U.K., a Ph.D. degree from University College, London, a D.Sc. degree from the University of Surrey, and is a Fellow of the Royal College of Pathologists. He has held posts as a Biochemist at University College Hospital, London (1974–1982); Lecturer at University College, London (1982–1984); Senior Lecturer (1984–1990); Reader (1990–1994); and Professor in the Department of Medicine at the University of Liverpool. Dr. Jackson was a member of the editorial board of the *British Journal of Nutrition,* (1988–1994). He currently serves on the editorial boards of *Basic and Applied Myology,* (1997–present); *Antioxidants and Redox Signalling,* (1999–present), and was Editor-in-Chief of *Clinical Science* (1997–1998). He is a Council Member of the International Society for Pathophysiology (1998–present). His current research funding sources include the Medical Research Council, Biotechnology and Biological Sciences Research Council, and the Wellcome Trust. Dr. Jackson's research interests include the role of antioxidant nutrients in regulation of cell viability and cellular responses to stress and whole body homeostasis of micronutrients.

# *Contributors*

**Steven A. Abrams, M.D.** USDA/ARS Children's Nutrition Research Center, Houston, TX, U.S.A.

**Claudio Cobelli, Ph.D.** Department of Electronics and Informatics, University of Padova, Padova, Italy.

**Helen M. Crews, Ph.D.** Ministry of Agriculture, Fisheries and Food, Central Science Laboratory, Sand Hutton, York, U.K.

**J. Dainty** Institute of Food Research, Norwich Research Park, Colney, Norfolk, U.K.

**Lena Davidsson, M.D.** Laboratory for Human Nutrition, Institute of Food Science, Swiss Federal Institute of Technology, Zürich, Switzerland.

**S.J. Fairweather-Tait** Institute of Food Research, Norwich Research Park, Colney, Norfolk, U.K.

**John W. Finley, Ph.D.** U.S. Department of Agriculture, Agricultural Research Service, Grand Forks Human Nutrition Research Center, Grand Forks, ND, U.S.A.

**T.E. Fox** Institute of Food Research, Norwich Research Park, Colney, Norfolk, U.K.

**R.S. Gibson, Ph.D.** Department of Human Nutrition, University of Otago, Dunedin, New Zealand.

**Marianne Hansen, Ph.D.** Research Department of Human Nutrition, The Royal Veterinary and Agricultural University, Frederiksberg, Denmark.

**L.J. Harvey** Institute of Food Research, Norwich Research Park, Colney, Norfolk, U.K.

**C. Hotz, Ph.D.** Program in International Nutrition, University of California, Davis, CA, U.S.A.

**Mats Isaksson, Ph.D.** Department of Radiation Physics, Göteborg University, Göteborg, Sweden.

**Catherine I.A. Jack, M.D.** Department of Geriatric Medicine, Broadgreen University Hospital Trust, Liverpool, U.K.

**Malcolm J. Jackson, D.Sc.** Department of Medicine, University of Liverpool, Liverpool, U.K.

**Morteza Janghorbani, Ph.D.** Center for Stable Isotope Research Inc., Chicago, IL, U.S.A.

**Nicola M. Lowe, Ph.D.** Department of Biological Sciences, University of Central Lancashire, Preston, U.K.

**Brittmarie Sandström, Ph.D.** Research Department of Human Nutrition, The Royal Veterinary and Agricultural University, Frederiksberg, Denmark.

**Alessandro Stevanato** Department of Electronics and Informatics, University of Padova, Padova, Italy.

**David M. Shames, M.D.** Department of Radiology, University of California, San Francisco, CA, U.S.A.

**B. Teucher** Institute of Food Research, Norwich Research Park, Colney, Norfolk, U.K.

**Gianna Toffolo, Ph.D.** Department of Electronics and Informatics, University of Padova, Padova, Italy.

**Judith R. Turnlund** Western Human Nutrition Research Center, USDA/ARS, University of California, Davis, CA, U.S.A.

**Leslie R. Woodhouse, Ph.D.** Western Human Nutrition Research Center, USDA/ARS, University of California, Davis, CA, U.S.A.

# Contents

**Chapter 1 Advances in Stable-isotope Methodology**
*Leslie R. Woodhouse and Steven A. Abrams*
1.1 History ... 1
1.1.1 First Use of Stable Isotopes with Humans — Deuterium and $^{15}N$ ... 2
1.1.2 Use of Mass Spectrometry for Mineral Stable-isotope Research ... 2
1.2 Using Stable Isotopes to Study Trace-element Metabolism ... 3
1.2.1 Advantages and Disadvantages ... 3
1.2.2 Stable-isotope Elements Available for Research ... 4
1.2.3 Instrumentation for Mineral Stable-isotope Research ... 8
1.2.3.1 Neutron Activation Analysis (NAA) ... 9
1.2.3.2 Gas Chromatography Mass Spectrometry (GC-MS) ... 9
1.2.3.3 Thermal Ionization Mass Spectrometry (TIMS) ... 9
1.2.3.4 Inductively Coupled Plasma Mass Spectrometry (ICP-MS) ... 10
1.2.3.5 Fast Atom Bombardment Mass Spectrometry (FAB-MS) ... 11
1.3 Stable-isotope Dosage, Preparation, and Administration ... 12
1.4 Practical Strategies for Conducting Stable-isotope Tracer Studies ... 14
1.4.1 Zinc ... 14
1.4.2 Iron ... 17
1.5 Appendix — Stable-isotope Suppliers ... 18
References ... 19

**Chapter 2 Advances in Radioisotope Methodology**
*Marianne Hansen, Mats Isaksson, and Brittmarie Sandström*
2.1 Introduction ... 23
2.2 Radioisotopes ... 24
2.3 Whole-body Counting Techniques ... 28
2.3.1 Whole-body Counting ... 28
2.3.2 Whole-body Counting Applications ... 29
2.3.2.1 Metabolism and Biological Turnover Rate ... 29
2.3.2.2 Absorption Studies ... 30
2.3.3 Equipment and Technological Development ... 31
2.4 Body Imaging Techniques ... 34
2.5 Indirect Measurements of Absorption or Metabolism ... 35
2.5.1 Tissue Retention ... 35
2.5.2 Urinary Excretion ... 36

2.5.3 Fecal Monitoring ..... 38
2.5.4 Equipment and Technological Development ..... 38
2.6 Conclusion ..... 39
References ..... 39

**Chapter 3 Tracer-to-tracee Ratio for Compartmental Modelling of Stable-isotope Tracer Data**

*Gianna Toffolo, David M. Shames, Alessandro Stevanato, and Claudio Cobelli*

3.1 Introduction ..... 43
3.2 Single-pool Tracer Kinetics and Measurement ..... 44
3.3 Tracer-to-tracee Ratio from Mass Spectrometry Measurements ..... 47
3.4 Multi-pool Tracer Kinetics and Measurement ..... 50
3.5 The Multiple Tracer Case ..... 52
3.6 A Test of the Endogenous-constant, Steady-state Assumption ..... 54
3.7 Software Tool: TTRM ..... 54
3.8 Conclusion ..... 56
References ..... 56

**Chapter 4 Methods for Analysis of Trace-element Absorption**

*S.J. Fairweather-Tait, T.E. Fox, L.J. Harvey, B. Teucher, and J. Dainty*

4.1 General Introduction ..... 60
4.1.1 Use of Isotopes ..... 60
4.1.2 Methods ..... 60
4.1.3 Definition of Absorption ..... 61
4.2 Iron ..... 62
4.2.1 Introduction ..... 62
4.2.2 Normalization of Iron Absorption Data ..... 62
4.2.3 Hemoglobin Incorporation ..... 63
4.2.4 Whole-body Counting ..... 64
4.2.5 Fecal Monitoring ..... 64
4.2.6 Plasma Appearance/Disappearance ..... 65
4.2.7 *In vitro* (Caco-2 Cells) ..... 65
4.2.8 Conclusion ..... 66
4.3 Copper ..... 66
4.3.1 Introduction ..... 66
4.3.2 Fecal Monitoring ..... 66
4.3.3 Plasma Appearance ..... 67
4.3.4 Whole-body Counting ..... 67
4.3.5 Conclusion ..... 68
4.4 Zinc ..... 68
4.4.1 Introduction ..... 68
4.4.2 Whole-body Counting ..... 68
4.4.3 Fecal Monitoring ..... 69
4.4.4 Urinary Monitoring ..... 70
4.4.5 Plasma Appearance/Disappearance ..... 70

4.4.6 Use of Simulation to Predict Absorption....71
4.4.7 Whole-gut Lavage Technique....71
4.4.8 *In vitro* (Caco-2 Cells)....72
4.4.9 Conclusion....72
4.5 Selenium....72
4.5.1 Introduction....72
4.5.2 Fecal Monitoring....74
4.5.3 Plasma Appearance/Disappearance....74
4.5.4 Whole-body Counting....75
4.5.5 Urinary Monitoring....75
4.5.6 Conclusion....76
References....76

**Chapter 5 Kinetic Studies of Whole-body Trace-element Metabolism**
*Nicola M. Lowe and Malcolm J. Jackson*
5.1 Introduction....81
5.2 General Considerations in Study Design....82
5.2.1 Isotope Dose....82
5.2.2 Sampling Strategy....82
5.2.3 Free-Living or Metabolic Unit....83
5.3 Compartmental Modelling....83
5.3.1 General Assumptions....84
5.4 Specific Examples of Isotope Turnover Studies....85
5.4.1 Zinc....85
5.4.2 Copper....86
5.4.3 Selenium....88
5.5 Conclusion....89
References....90

**Chapter 6 Stable-isotope Methods for the Investigation of Iron Metabolism in Man**
*Morteza Janghorbani*
6.1 Introduction....93
6.2 Iron Metabolism in Relation to the Design of Stable-isotope Protocols....94
6.3 Feasibility Issues....95
6.4 Analytical Methods....99
6.4.1 Neutron Activation Analysis....99
6.4.2 Mass Spectrometry....100
6.4.3 Summary of Current Analytical Capabilities....101
6.5 Selected Applications....102
6.5.1 Relationship between Mucosal Absorption and Hemoglobin Incorporation of Dietary Iron....102
6.5.2 Issues of Dietary Availability of Iron....103
6.6 Conclusion....104
References....105

**Chapter 7 Use of Isotopes in the Assessment of Zinc Status**

*Malcolm J. Jackson and Nicola M. Lowe*


7.1 Introduction ..... 109
7.2 Isotopic Techniques ..... 111
7.2.1 Short-term Two-compartment Model ..... 112
7.2.2 Simplified Techniques to Measure the Exchangeable Zinc Pool ..... 113
7.3 Conclusion ..... 113
References ..... 114


**Chapter 8 Copper Status and Metabolism Studied with Isotopic Tracers**

*Judith R. Turnlund*


8.1 Introduction ..... 117
8.2 Background ..... 118
8.3 Copper Status ..... 118
8.4 Isotopic Tracers ..... 119
8.4.1 Radioactive Tracers ..... 119
8.4.2 Stable-isotope Tracers ..... 120
8.4.2.1 Methods of Stable-isotope Analysis ..... 120
8.4.2.1.1 Neuron Activation Analysis ..... 120
8.4.2.1.2 Electron Impact Mass Spectrometry and Gas Chromatography Mass Spectrometry ..... 120
8.4.2.1.3 Thermal Ionization Mass Spectrometry ..... 121
8.4.2.1.4 Inductively Coupled Plasma Mass Spectrometry ..... 121
8.4.2.2 Multiple Stable-isotope Labelling ..... 121
8.4.2.3 Studies Using Isotopic Tracers of Copper ..... 122
8.5 Conclusion ..... 123
References ..... 123


**Chapter 9 Use of Stable Isotopes of Selenium to Investigate Selenium Status**

*Helen M. Crews*


9.1 Introduction ..... 130
9.2 Dietary Selenium and Its Metabolism ..... 130
9.2.1 Sources and Daily Intakes ..... 130
9.2.2 Chemical Form and Bioavailability ..... 131
9.2.3 Metabolism of Selenium ..... 132
9.3 The Role of Selenium in the Body ..... 133
9.3.1 Selenium and Disease ..... 133
9.3.1.1 Selenium Deficiency and Disease ..... 133
9.3.1.2 Selenium and Cancer ..... 134
9.3.2 Selenoproteins ..... 134
9.3.2.1 Intracellular Glutathione Peroxidases (EC 1.11.1.9.) ..... 135
9.3.2.1.1 Cellular (Cystolic) GSHpx ..... 135

9.3.2.1.2 Phospholipid Hydroperoxide GSHpx ....... 135
9.3.2.1.3 Gastrointestinal GSHpx ....... 136
9.3.2.2 Extracellular GSHpx ....... 136
9.3.2.2.1 Plasma GSHpx ....... 136
9.3.2.3 Iodothyronine Deiodinases (EC 3.8.1.4.) ....... 136
9.3.2.4 Thioredoxin Reductase (EC 1.6.4.5.) ....... 136
9.3.2.5 Selenium-binding Protein ....... 137
9.3.2.6 Others ....... 137
9.4 Assessment of Selenium Status and Use of Stable Isotopes ....... 137
9.4.1 Status Assays ....... 137
9.4.2 Analytical Aspects ....... 138
9.4.2.1 Assays for GSHpx Activity ....... 138
9.4.2.2 Measurement of Selenium Isotopes ....... 139
9.4.3 Modelling of Selenium Body Pools ....... 140
9.4.4 Stable-isotope Studies with Low-to-medium Selenium Intakes ....... 143
9.4.5 Stable-isotope Studies with High Selenium Intakes ....... 144
9.5 Conclusion ....... 145
References ....... 146

**Chapter 10 Use of Isotopes for Studies with Manganese, Chromium, and Molybdenum**

*John W. Finley*

10.1 Manganese ....... 152
10.1.1 Introduction ....... 152
10.1.2 Manganese Biochemistry ....... 152
10.1.3 Radioactive Isotopes of Manganese and Studies of Manganese Essentiality ....... 153
10.1.3.1 Studies with Laboratory Animals and Cultured Cells ....... 153
10.1.3.2 Distribution and Retention of Radioactive Manganese in Humans ....... 154
10.1.3.3 Radioactive Methods of Determining Apparent Manganese Absorption in Humans ....... 155
10.1.3.4 Radioactive Methods for Determining True Manganese Absorption ....... 156
10.1.3.5 The Use of Radioisotopes to Study Manganese/Iron Interactions ....... 158
10.2 Chromium ....... 159
10.2.1 Introduction ....... 159
10.2.2 Chemistry and Biochemistry ....... 160
10.2.3 Radioactive Chromium in Human Studies ....... 160
10.2.3.1 Nutritional Studies with $^{51}Cr$ ....... 160
10.2.3.2 Stable Isotopes of Chromium in Human Studies ....... 161
10.3 Molybdenum ....... 161
10.3.1 Chemistry and Biochemistry ....... 161
10.3.2 Radioactive Isotopes of Molybdenum in Human Studies ....... 161

10.3.3 Stable Isotopes of Molybdenum in Human Studies ................ 162
10.4 Summary ................ 162
References ................ 163

**Chapter 11 Trace-element Studies in Infants and Pregnant or Lactating Women**

*Lena Davidsson*

11.1 Introduction ................ 167
11.2 Iron ................ 170
11.2.1 Methodology ................ 170
11.2.2 Erythrocyte Incorporation and Iron Absorption ................ 173
11.2.2.1 Studies in Premature Infants ................ 173
11.2.2.2 Studies in Term Infants ................ 174
11.2.2.2.1 Human Milk and Infant Formula ................ 174
11.2.2.2.2 Complementary Foods ................ 176
11.2.2.2.3 Iron Supplements ................ 177
11.2.2.3 Studies in Pregnant Women ................ 177
11.3 Zinc ................ 178
11.4 Zinc and Copper ................ 180
11.5 Selenium ................ 181
11.6 Chromium ................ 182
11.7 Conclusion ................ 183
References ................ 183

**Chapter 12 Stable-isotope Studies in the Elderly**

*Catherine I.A. Jack, Nicola M. Lowe, and Malcolm J. Jackson*

12.1 Introduction ................ 187
12.2 Practicalities of Working with Elderly Subjects ................ 188
12.3 Ethical Considerations ................ 188
12.4 Examples of Stable-isotope Studies in the Elderly ................ 189
12.4.1 Zinc Homeostasis in the Elderly ................ 189
12.4.2 Copper Homeostasis in the Elderly ................ 189
12.5 Selenium Status of the Elderly ................ 190
12.6 Conclusion ................ 190
Acknowledgments ................ 190
References ................ 191

**Chapter 13 Applications of Trace-element Studies in Developing Countries: Practical and Technical Aspects**

*R.S. Gibson and C. Hotz*

13.1 Introduction ................ 194

13.2 Applications of Isotope Studies in Developing Countries.................. 195
13.2.1 Supplementation .................................................................... 195
13.2.2 Fortification ........................................................................... 197
13.2.3 Dietary Strategies .................................................................. 198
13.3 Practical Aspects of Implementing Isotope Studies in Developing Countries ........................................................................ 199
13.3.1 Securing Support within the Country at the National and Community Level .................................................................... 199
13.3.2 Selecting the Study Design ................................................... 200
13.3.3 Assessing the Nutritional and Health Status of the Study Participants ...................................................... 201
13.3.4 Assessing Levels of Trace Elements and Absorption Modifiers in the Habitual Diets of Study Participants ............. 203
13.3.4.1 Assessing Food Intakes ................................................... 203
13.3.4.2 Compiling a Local Food Composition Table for Use in a Developing Country......................................... 204
13.3.4.3 Assessing Intakes of Trace Elements and Absorption Modifiers in Habitual Diets...................... 204
13.3.4.4 Assessing Nutrient Intakes during the Metabolic Study ............................................................ 205
13.4 Technical Aspects of Implementing Isotope Studies in Developing Countries ........................................................................ 206
13.4.1 Considerations When Selecting the Isotopic Technique .......... 207
13.4.1.1 Fecal Monitoring ............................................................. 207
13.4.1.2 Urinary Monitoring ......................................................... 208
13.4.1.3 Tissue Retention .............................................................. 209
13.4.1.4 Plasma Tolerance Curves and Plasma Deconvolution .. 209
13.4.2 Collecting, Preparing, and Processing the Metabolic Samples for Analysis of Native Trace Elements and Isotopic Enrichment .. 210
13.4.2.1 Fecal Samples .................................................................. 210
13.4.2.2 Urine Samples .................................................................. 211
13.4.2.3 Blood Samples ................................................................. 211
13.5 Conclusion ........................................................................................ 212
References ................................................................................................ 212

Index ........................................................................................................ 217

# 1

# *Advances in Stable-isotope Methodology*

**Leslie R. Woodhouse and Steven A. Abrams**

**CONTENTS**

1.1 History ........ 1
1.1.1 First Use of Stable Isotopes with Humans — Deuterium and $^{15}N$ ........ 2
1.1.2 Use of Mass Spectrometry for Mineral Stable-isotope Research ... 2
1.2 Using Stable Isotopes to Study Trace-element Metabolism ........ 3
1.2.1 Advantages and Disadvantages ........ 3
1.2.2 Stable-isotope Elements Available for Research ........ 4
1.2.3 Instrumentation for Mineral Stable-isotope Research ........ 8
1.2.3.1 Neutron Activation Analysis (NAA) ........ 9
1.2.3.2 Gas Chromatography Mass Spectrometry (GC-MS) ..... 9
1.2.3.3 Thermal Ionization Mass Spectrometry (TIMS) ........ 9
1.2.3.4 Inductively Coupled Plasma Mass Spectrometry (ICP-MS) ........ 10
1.2.3.5 Fast Atom Bombardment Mass Spectrometry (FAB-MS) ........ 11
1.3 Stable-isotope Dosage, Preparation, and Administration ........ 12
1.4 Practical Strategies for Conducting Stable-isotope Tracer Studies ....... 14
1.4.1 Zinc ........ 14
1.4.2 Iron ........ 17
1.5 Appendix — Stable-isotope Suppliers ........ 18
References ........ 19

## 1.1 History

Due to the rapid advances in mass spectrometry techniques over the last 20 years, a steady growth in the application of stable isotope use to study human mineral and trace-element metabolism has occurred. The most frequent

0-8493-8730-2/00/$0.00+$.50

application of stable isotopes of the mineral elements in studies of nutrition and metabolism has been to determine dietary mineral availability or absorption. With improved analytical precision, (mainly due to improved instrumentation) the versatility of stable-isotope tracer techniques has increased to include measures of endogenous excretion, and kinetic measures of mineral turnover and body pools, resulting from compartmental modelling. Several relatively recent review articles are available regarding the use of stable-isotope technology for trace mineral studies in humans, and older review articles are important historically for understanding the advances that have occurred in this field.[1–5]

### 1.1.1 First Use of Stable Isotopes with Humans — Deuterium and $^{15}N$

Stable isotopes were used in metabolic research prior to the use of radioactive isotopes. The first stable-isotopic tracer study was reported in 1935 by Schoenheimer and Rittenberg,[6] who used deuterium, the heavy isotope of hydrogen, to study intermediary metabolism in laboratory animals and humans. $^{15}N$ became available shortly thereafter.[4] The first mineral isotopes to be used as tracers were radioactive isotopes. Radioactive iron was first used in humans in 1942, and other radioactive mineral studies in humans using copper, calcium, zinc, magnesium, molybdenum, and selenium occurred between 1947 and 1970.[4] Due to the risks associated with radiation exposure, and the limitations in metabolic research that came about as a result of the restricted use of radioactive isotopes in most human populations, the exploration of stable isotopes for human mineral metabolic research increased.

### 1.1.2 Use of Mass Spectrometry for Mineral Stable-isotope Research

The first publication describing the use of a stable-isotope tracer in a human metabolic study was published in 1963.[4] An enriched stable isotope of iron, Fe-58, was injected into men in order to determine the plasma clearance of the stable isotope compared to the radioactive iron tracer, $^{59}Fe$. Through 1979, many more stable-isotope experiments were published with mineral stable-isotope tracers (Ca, Cr, Zn, Cu, Fe, Mg); all of these early studies used the technique of neutron activation analysis (NAA) to measure the isotopes.[4] At the same time, all of the stable-isotope analysis of the lower-mass, non-mineral elements was done using mass spectrometry techniques. A 1979 publication that described the use of electron impact mass spectrometry to determine $^{26}Mg$ enrichment marked the beginning of the current mass spectrometry era for mineral stable-isotope research.[7] Further instrumentation advances, including FAB-MS (fast atom bombardment mass spectrometry), TIMS (thermal ionization mass spectrometry), ICP-MS (inductively coupled plasma mass spectrometry), as well as the high resolution and magnetic sector ICP-MS instruments, have accelerated the field of mineral stable-isotope research. This has enabled many more researchers to become involved in the field to address the multitude of complex questions in the area of trace-element metabolism in humans and animals.

## 1.2 Using Stable Isotopes to Study Trace-element Metabolism

### 1.2.1 Advantages and Disadvantages

There are advantages and disadvantages with the use of stable isotopes in the study of trace-element metabolism which must be considered when designing experimental protocols utilizing stable isotopes. It is important to note nomenclature used to describe these stable-isotope minerals. An enriched isotope, when obtained from the supplier, is always "contaminated" with other stable isotopes of the same element, which are also considered tracers in the experiment and need to be considered in the calculations of isotope enrichment. To distinguish between a pure stable isotope and a tracer, different notations should be used. For example, Zn-70 designates the enriched isotope as purchased from the supplier, while $^{70}$Zn is the standard notation for the pure isotope.[8]

The primary advantage of stable-isotope minerals (and radioisotope minerals) is that they can be used to trace mineral metabolic fate. The important nutritional questions answered include: bioavailability of the mineral with or without specific foods; dose effects; trace element interactions; and mineral absorption.

The most important advantage of the use of stable isotopes is the fact that the use of non-radioactive labels increases the safety of the technique in all populations as well as allowing populations such as infants and pregnant women to be studied. Also, because there is no isotopic decay, the element can be traced in the body for a long period of time (as long as there is sufficient enrichment) and the samples collected can be stored indefinitely without loss of "signal." There are some elements that have limited use for study with radioactive tracers, due to short half-lives ($^{28}$Mg, 21 hours, and $^{67}$Cu, 62 hours). These elements have suitable stable isotopes that enable more appropriate metabolic studies.

Another advantage of the mineral stable isotopes is the number of isotopes available for a particular mineral. Most of the minerals have isotopes of relatively low natural abundance, which enables multiple isotopes of the same element to be used simultaneously for study, as well as multiple isotopes of different elements. Because stable isotopes are naturally present in the body, the natural isotopic abundance and the degree of required enrichment in the biological samples to be measured are very important considerations when planning mineral studies. The tracer of choice is the isotope of the least abundant naturally occurring isotopes, which would allow for much less of the tracer to be used for isotope administration, either orally or intravenously.

When the analysis of stable-isotope elements is based on mass, isotopes of the same element, or of different elements, will not interfere with one another in the analysis. However, mono-isotopic elements, including F, P and Mn cannot be studied using stable isotopes. Table 1.1 contains a list of the essential minerals and their isotopic distribution, including those trace elements with undefined requirements that may be nutritionally relevant. Depending on the published source used for the isotopic distribution, the numbers will vary slightly.[9–14] Although the natural variability (fractionation) of the mineral isotopes in nature is extremely small, slight differences in measured natural abundances occur, due to the techniques utilized for their measurements.

### 1.2.2 Stable-isotope Elements Available for Research

Stable isotopes may be relatively expensive, with the cost depending on the natural abundance, enrichment level, and availability, as well as the country of origin, supplier, and quantity ordered. Because there are no disposal costs related to their use, however, it is not always true that isotope costs are prohibitively more expensive for stable compared to radioactive tracers. Table 1.2 is a listing of currently available stable-isotope elements, the enrichment ranges available, and approximate costs. These prices are from Oak Ridge National Laboratories and are generally quoted higher than quotes available from other isotope suppliers. This listing is subject to change, but can give the investigator a "ball park" idea of comparative costs involved. Isotope suppliers work very closely with investigators to supply isotopes at varying levels of enrichment from 1% to 99%+, and establish prices based on quantity, enrichment grade, and customer commitment. Many stable isotopes can be produced with short-term notice, and most companies have highly enriched isotopes in stock. Appendix I is a listing of many of the companies that currently market or produce stable isotopes.

A potential limiting factor in mineral stable-isotope studies is the lack of available sites for their analysis. Most facilities with the capacity for analyzing these samples are associated with geology research facilities. However, this situation is also improving. The availability of more techniques and newer equipment such as advanced TIMS and ICP mass spectrometers, and the willingness of non-nutrition laboratories to collaborate in these research projects, have led to an increased availability of analytical sites.

The substantial sample preparation needed prior to isotope analysis has also been limiting; nevertheless, these techniques are well described and it is possible that some newer analytical techniques such as magnetic sector ICP-MS will not need extensive sample preparation.

Another issue concerning stable-isotope studies is that they are not necessarily used as true "tracers" as with a radioactive tracer. All the stable isotopes occur in nature, so they need to be studied using amounts greater than their natural abundance in order to detect enrichment levels. For example,

**TABLE 1.1**

Isotopic Composition of Minerals Essential to Humans

| Mineral | Isotopic Weight | Abundance[a] |
|---|---|---|
| *Macrominerals with Established RDA Values* | | |
| Calcium | 40 | 96.941 |
| | 42 | 0.647 |
| | 43 | 0.135 |
| | 44 | 2.086 |
| | 46 | 0.004 |
| | 48 | 0.187 |
| Magnesium | 24 | 78.992 |
| | 25 | 10.003 |
| | 26 | 11.005 |
| *Trace Elements with Established RDA Values* | | |
| Iodine | 127 | 100 |
| Iron | 54 | 5.810 |
| | 56 | 91.750 |
| | 57 | 2.150 |
| | 58 | 0.290 |
| Selenium | 74 | 0.889 |
| | 76 | 9.366 |
| | 77 | 7.635 |
| | 78 | 23.772 |
| | 80 | 49.607 |
| | 82 | 8.731 |
| Zinc | 64 | 48.630 |
| | 66 | 27.900 |
| | 67 | 4.100 |
| | 68 | 18.750 |
| | 70 | 0.620 |
| *Trace Elements with ESADDI* | | |
| Chromium | 50 | 4.345 |
| | 52 | 83.790 |
| | 53 | 9.501 |
| | 54 | 2.365 |
| Copper | 63 | 69.174 |
| | 65 | 30.826 |
| Fluoride | 19 | 100 |
| Manganese | 55 | 100 |
| Molybdenum | 92 | 14.836 |
| | 94 | 9.247 |
| | 95 | 15.920 |
| | 96 | 16.676 |
| | 97 | 9.555 |
| | 98 | 24.133 |
| | 100 | 9.634 |

**TABLE 1.1 (*continued*)**

Isotopic Composition of Minerals Essential to Humans

| Mineral | Isotopic Weight | Abundance[a] |
|---|---|---|
| *Trace Elements with Undefined Requirements* | | |
| Arsenic | 75 | 100 |
| Boron | 10 | 19.820 |
| | 11 | 80.180 |
| Bromine | 79 | 50.686 |
| | 81 | 49.314 |
| Lead | 204 | 1.425 |
| | 206 | 24.145 |
| | 207 | 22.083 |
| | 208 | 52.348 |
| Nickel | 58 | 68.077 |
| | 60 | 26.223 |
| | 61 | 1.140 |
| | 62 | 3.635 |
| | 64 | 0.926 |
| Silicon | 28 | 92.229 |
| | 29 | 4.670 |
| | 30 | 3.101 |
| Tin | 112 | 0.973 |
| | 114 | 0.652 |
| | 115 | 0.339 |
| | 116 | 14.537 |
| | 117 | 7.676 |
| | 118 | 24.225 |
| | 119 | 8.586 |
| | 120 | 32.595 |
| | 122 | 4.629 |
| | 124 | 5.789 |
| Vanadium | 50 | 0.250 |
| | 51 | 99.750 |

[a] Source: References 15 and 16.

*Note:* RDA: Recommended Dietary Allowance; ESADDI: Estimated Safe and Adequate Daily Dietary Intakes.

with the element Cu, the $^{63}Cu$ and $^{65}Cu$ occur naturally at 69.2% and 30.8%. In order to use the $^{65}Cu$ as a tracer, a large amount of a highly enriched preparation of Cu-65 would need to be used to see sufficient enrichment levels above the high "background" of the naturally occurring $^{65}Cu$. This limits its application for metabolic studies (especially for intravenous use) because large, non-physiological quantities of the isotope would be necessary, which may perturb mineral metabolism in the subject. Generally, if an isotope used as a tracer is greater than five percent at natural abundance, a relatively high dose of isotope needs to be administered in order to achieve measurable enrichment in the biological samples. This dose may represent a significant fraction of the exchangeable mineral pool, and therefore may not be functioning as a true tracer.[2] Ideally, intravenous tracers should be kept at levels of less

**TABLE 1.2**

Commercially Available Stable Isotopes

| Mineral | Isotopic Weight | Enrichment, % | Cost per mg of Element (U.S. $) |
|---|---|---|---|
| Calcium | 40 | 99+ | 1 |
| | 42 | 93,94 | 75 |
| | 43 | 84 | 450 |
| | 44 | 98 | 30 |
| | 46 | 31 | 4150 |
| | 48 | 98 | 280 |
| Magnesium | 24 | 99+ | 3 |
| | 25 | 98 | 15 |
| | 26 | 99+ | 11 |
| Iron | 54 | 97 | 20 |
| | 56 | 99+ | 1 |
| | 57 | 92–95 | 7–14 |
| | 58 | 82 | 200 |
| Selenium | 74 | 78 | 760 |
| | 76 | 97 | 30 |
| | 77 | 94 | 34 |
| | 78 | 99+ | 12 |
| | 80 | 99+ | 5 |
| | 82 | 97 | 37 |
| Zinc | 64 | 99+ (also <1) | 9 |
| | 66 | 99+ | 7 |
| | 67 | 94 | 50 |
| | 68 | 99+ | 5 |
| | 70 | 85–90 | 440 |
| Chromium | 50 | 97 | 80 |
| | 52 | 99+ | 3 |
| | 53 | 96 | 35 |
| | 54 | 95 | 180 |
| Copper | 63 | 99+ | 2 |
| | 65 | 99+ | 5 |
| Molybdenum | 92 | 97 | 4 |
| | 94 | 92 | 6 |
| | 95 | 94 | 4 |
| | 96 | 97 | 3 |
| | 97 | 94 | 6 |
| | 98 | 98 | 3 |
| | 100 | 98 | 6 |
| Boron | 10 | 92–99+ | 80 |
| | 11 | 95 | 10 |
| Bromine | 79 | 99+ | 12 |
| | 81 | 99+ | 13 |
| Lead | 204 | 71–99+ | 70–120 |
| | 206 | 99+ | 5 |
| | 207 | 93 | 5 |
| | 208 | 99+ | 2 |
| Nickel | 58 | 99+ | 1 |
| | 60 | 99+ | 2 |
| | 61 | 99+ | 70 |
| | 62 | 99+ | 20 |

**TABLE 1.2 (*continued*)**

Commercially Available Stable Isotopes

| Mineral | Isotopic Weight | Enrichment, % | Cost per mg of Element (U.S. $) |
|---|---|---|---|
| | 64 | 99+ | 50 |
| Silicon | 28 | 99+ | 4 |
| | 29 | 96 | 75 |
| | 30 | 96 | 160 |
| Chloride | 35 | 99+ | 8 |
| | 37 | 98 | 30 |
| Potassium | 39 | 99+ | 9 |
| | 40 | 3 | 25 |
| | 41 | 99+ | 160 |

*Source:* Oak Ridge National Laboratories.

than or equal to three percent of the elemental/molecular pool into which they are administered[17] so as not to perturb the homeostasis of the subject.

Unlike radioisotopes used in metabolic studies, stable isotopes cannot be detected *in vivo*, thus limiting the physiological location of samples that can be assessed. The sites that can readily be measured in human metabolic studies are limited to blood, excreta (feces and urine), saliva, and milk (during lactation). Under specialized circumstances, gastric lavage and sampling can also be used to assess mineral absorption and metabolism.[18,19]

### 1.2.3 Instrumentation for Mineral Stable-isotope Research

Although several analytical approaches have been used for the isotopic analysis of inorganic elements, mass spectrometry is currently the principal analytical technique utilized.[20,21] Neutron activation analysis (NAA), as mentioned previously, was the first analytical technique utilized with mineral stable isotopes. Gas chromatography mass spectrometry (GC-MS) is used for the analysis of volatile metal chelates, so the analysis is therefore limited to those metals that form volatile chelates. More recently, FAB-MS, ICP-MS, and TIMS are the three methods most widely used, with ICP-MS and TIMS as the primary analytical instruments for stable isotope research with trace elements. These three MS instrumentation methods vary with respect to analysis time per sample and precision attained, but are quite similar with respect to sample size needed for analysis, sample preparation, and dedicated operator skill. The instrumentation costs are quite different; quadrupole ICP-MS is currently the most affordable instrument. Table 1.3 shows approximate initial investment costs associated with the three main MS techniques. Approximate annual running costs associated with consumables, such as gas use and disposables, is approximately \$10,000 to \$20,000 per year, with extra costs for potential service contracts. The marked improvement in analytical technology with stable isotopes for nutritionally important minerals has accelerated the number of studies conducted and the speed at which they are completed.

**TABLE 1.3**

Approximate Initial Investment Costs

| Instrument | Approximate Cost (in year 2000 U.S. $) |
|---|---|
| FAB-MS | 700,000 |
| TIMS: | |
| Quadrupole | No longer in production |
| Magnetic Sector | 600,000 |
| ICP-MS: | |
| Quadrupole | 75,000–200,000 |
| High Resolution Magnetic Sector | 350,000 |
| High Resolution Multiple Collector | 750,000 |

#### *1.2.3.1 Neutron Activation Analysis (NAA)*

NAA is primarily used for total element analysis, but can be used to determine isotopic enrichment.[3] NAA for determination of stable-isotope enrichment is based on the interaction of thermal neutrons from a nuclear reactor with the nuclei of the stable isotopes. The nuclear transformation results in production of a radioactive nuclide that emits radiation measured with a Ge(Li) detector coupled to a multichannel analyzer.[3] Some isotopes do not result in the production of the radioactive nuclide and cannot be measured with NAA. Other elements, such as $^{70}Zn$, require extensive pre- and post-radiochemical separation, resulting in fairly low precision of about 5%. On the other hand, for iron, which requires little sample preparation, precision is about 1%.

#### *1.2.3.2 Gas Chromatography Mass Spectrometry (GC-MS)*

GC-MS analysis for metal isotope ratios has the advantage of being the most widely available and least expensive technique of all MS analyses.[20] A volatile chelate of the metal is formed and introduced to the MS as an eluent from a gas chromatograph. The technique has been used for chromium, selenium, nickel, platinum, mercury, cobalt, lead, copper, and mercury; there are no satisfactory chelates for iron or zinc. Precision is marginal at 1 to 2%. The main experimental difficulty with GC-MS is the selection of a chelating agent and ion source conditions (either electron ionization or chemical ionization), to ensure that there are no memory effects between samples. Aggarwal et al.[22–24] have published several manuscripts describing work with GC-MS and metal isotopes, and at this time it appears that the use of GC-MS for metabolic studies is best for chromium and selenium studies.[20]

#### *1.2.3.3 Thermal Ionization Mass Spectrometry (TIMS)*

TIMS, initially developed for geologic research, is considered the method of choice for stable-isotope analysis due to its high precision. Both quadrupole-separating TIMS and magnetic-sector TIMS have been commercially produced.

Magnetic-sector machines separate masses in a magnetic field; quadrupole machines separate masses by applying alternating and constant voltages to parallel rods. The magnetic sector instruments yield the best precision (<0.1%). A major disadvantage of TIMS analysis is sample throughput: it is difficult to analyze more than 10–15 samples per day for trace element analysis. TIMS instrumentation is very expensive and usually requires a dedicated operator; nonetheless, TIMS will continue to play a major role in stable-isotope studies of mineral nutrition and metabolism due to its extremely high accuracy and precision.

#### 1.2.3.4 Inductively Coupled Plasma Mass Spectrometry (ICP-MS)

ICP-MS, the most recent of the MS techniques, is the most widely used metal isotope analysis technique, and is an instrument specifically designed for trace-element quantitation. The first commercial ICP-MS system was introduced to the marketplace in 1983, and there are currently over a dozen companies that manufacture ICP-MS instruments.[25] The first report of an ICP-MS used for trace-element quantitation occurred in 1975;[26] one of the first stable-isotope studies done with ICP-MS, investigating the incorporation of Fe-58 in erythrocytes of children, was reported by Janghorbani et al in 1986.[27] Samples are introduced through a nebulizer into a high temperature argon plasma produced by electrical discharge, where the solids are volatilized and ionized. The plasma is sampled at atmospheric pressure through a differentially pumped interface, and ions are usually separated by mass with a quadrupole mass spectrometer.

In the mid-1990s, commercially produced ICP-MS instruments which separate ions using a magnetic field (high resolution magnetic sector ICP-MS) became available through several manufacturers. These machines feature very high resolution with claims of analytical precision close to that achieved using TIMS as well as very rapid analysis times (about 5 minutes per sample).[28] Also very new to the market, and currently very expensive, are the multicollector ICP-MS instruments, which produce high precision isotope ratio measurements.[29]

Depending on the mineral, sample preparation is important for ICP-MS analysis, as there are interferences, and some elements cannot be analyzed easily due to these interferences at specific masses. Some of the interferences include interelement (isobaric) spectral overlaps; polyatomic interferences resulting from background species from the argon, water, and air; and molecular species derived from the analytes in the sample or the sample matrix.[30] Because of the wide variety of matrix effects in ICP-MS, it is important for the investigator to be aware of these possibilities because many of the spectral interferences in ICP-MS are specific to particular matrices and operating conditions.[31] Some of the software packages for operation of the ICP-MS contain a spectral interference database, and there are also database programs available for use, e.g., MS InterView.[30] The newer quadrupole ICP-MS instruments are very sensitive, especially when coupled with improved sample introduction devices, such as ultrasonic nebulizers. The high resolution magnetic sector

**TABLE 1.4**

Commercial ICP-MS Instruments

| Instrument Name | Manufacturer |
|---|---|
| *Quadrupole Instruments* | |
| Elan 6100 | Perkin Elmer/Sciex Corp. |
| HP 4500 | Hewlett-Packard Corp./Yokogawa Analytical |
| POEMS II | Thermo Jarrell Ash Corporation |
| PlasmaQuad 3, PQExcell | VG Elemental |
| UltraMass 700 | Varian Analytical Instruments |
| SPQ 9000 | Seiko Instruments |
| SpectroMass 2000 | Spectro Analytical Instruments |
| Platform-ICP | Micromass U.K., Ltd. |
| ICPM-8300 | Shimadzu Scientific Instruments |
| *High-Resolution Magnetic Sector Instruments* | |
| Element 2 | Finnigan MAT Corp. |
| JMS-Plasmax 2 | JEOL Inc. |
| Plasma Trace 2 | Micromass U.K., Ltd. |
| VG Axiom SC | VG Elemental |
| *MultiCollector Magnetic Sector Instruments* | |
| Neptune | Finnigan MAT |
| Nu Plasma, Nu 1700 | Nu Instruments, Ltd. |
| IsoProbe | Micromass U.K., Ltd. |
| Axiom, Plasma 54 | VG Elemental |

*Source:* Reference 32.

instruments have even lower limits of detection, and much less interference due to the high mass resolution. The high resolution ICP-MS instruments spectrally separate interfering masses by coupling the Ar ICP source to a high-resolution mass spectrometer. For example, the high-resolution magnetic sector instrument can resolve the mass signal of $^{56}Fe$ from ArO+. This cannot be done with quadrupole mass analyzers, which is the main reason iron isotopes are difficult to analyze with quadrupole ICP-MS. Furthermore, very limited sample preparation may be necessary using these instruments.[28] These instruments are also fairly easy to operate, due to the user-friendly software and software-run instrumentation controls. Table 1.4 is a partial listing of commercially available ICP-MS instruments.

#### *1.2.3.5 Fast Atom Bombardment Mass Spectrometry (FAB-MS)*

FAB-MS is well known for the analysis of large, labile polar molecules, but has also been shown to be useful for analysis of stable-mineral isotopes, especially zinc.[33,34] In FAB-MS, samples are bombarded with argon or xenon atoms, and the sputtered charged species are separated in the mass analyzer. Analysis time is about 30 minutes per sample, with precision of approximately 1%, and down to 0.2 to 0.6% for zinc isotopes.[33]

## 1.3 Stable-isotope Dosage, Preparation, and Administration

The dose of the isotope to be used in a study primarily depends on the natural abundances of the enriched isotopes and the reference isotope. The size and physiological status of the subject also need to be considered. The expected sample enrichment can be estimated by knowing the approximate mineral content of the samples to be analyzed, the expected absorption and retention of the mineral, and the sampling time post-enrichment. The precision of the isotope ratio measurement also needs to be considered here.

The best tracer is the isotope that is lowest in natural abundance, as less of the mineral needs to be administered. If a dual isotope tracer experiment is being designed, the intravenous tracer is usually the isotope of lowest natural abundance, and the oral tracer is the isotope of second lowest abundance. For example, in a typical zinc dual-isotope tracer study, Zn-70 is often infused intravenously at levels of 0.3 to 1.0 mg. The next lowest abundant isotope is $^{67}$Zn, which occurs naturally at 4.1%. Since $^{70}$Zn occurs naturally at 0.62%, in order to achieve the same level of enrichment in the biological samples collected during the zinc study, 6.6-fold more Zn-67 would have to be infused (2.0 to 6.6 mg), since $^{67}$Zn is 6.6 times more abundant than $^{70}$Zn. Infusing such a large amount of zinc into the circulation may perturb zinc homeostasis, as there is only about 3 mg of total zinc circulating in the plasma. In general, the higher the natural abundance of the element, the greater the quantity of isotope which must be administered in order to detect an enrichment. In specific circumstances involving trace minerals, however, it may be optimal to give the least abundant isotope orally. For example, in studies of infants, such as breast-feeding babies, low concentrations of minerals are being traced. Using Fe-58 or Zn-70 orally in such cases allows the oral isotope dose to represent the smallest possible fraction of the dietary intake.[35]

Knowing the expected mineral content of the samples to be analyzed, as well as the expected absorption of the mineral, is helpful for determining dosage levels. For example, iron absorption can vary from 1% up to 90%, depending on the iron status of the subject, or the method of isotope administration (i.e., with food, with water, or with ascorbic acid). Zinc absorption can also vary depending on dietary zinc levels, and selenium absorption is usually very high. These estimations of absorption will help determine isotope enrichment of the feces. Urinary excretion of the element also needs to be considered because urinary iron is extremely low, and zinc is also relatively low, while urinary selenium is quite high. Table 1.5 shows trace-mineral, stable-isotope doses frequently used in nutritional studies.[2,36–38] Pediatric studies generally use lower dosages for trace minerals; although, in the case of calcium, because the proportion of bone mass that is highly turning over is maximum in early puberty, the total dose of isotope used in young adolescents is frequently greater than that needed for adults.[2]

**TABLE 1.5**

Trace Mineral Stable Isotopes Frequently Used in Nutritional Research with Humans

| Isotope | Natural Abundance, % | Typical Dose Ranges[a] |
|---|---|---|
| Cu-65 | 30.83 | 0.2–2.5 mg p.o.<br>Infant: 0.05–0.2 mg p.o. |
| Fe-57 | 2.15 | 5–15 mg p.o.<br>Infant: 2–4 mg p.o. |
| Fe-58 | 0.29 | 1–3 mg po, 0.2–0.4 mg IV<br>Infant: 0.2–0.5 mg p.o. |
| Mo-94 | 9.25 | 35 ug IV; 100 ug p.o. |
| Mo-97 | 9.56 | 35 ug IV; 100 ug p.o. |
| Mo-100 | 9.63 | 35 ug IV; 100 ug p.o. |
| Zn-67 | 4.10 | 1–3 mg p.o., 1 mg IV<br>Infant: 0.5 mg IV |
| Zn-68 | 18.75 | 3 mg p.o. |
| Zn-70 | 0.62 | 0.2–0.5 mg IV<br>Infant: 0.2–0.3 mg p.o. |

[a] IV: intravenous dose, p.o.: oral dose.

The length of time enrichment needs to be determined also is an important consideration regarding isotope dosage. If the study is a 10-hour plasma disappearance study, the intravenous dose can be quite small. If endogenous excretion of a mineral needs to be determined, fecal collections are usually required for 7 to 14 days after isotope administration. Also, gastrointestinal transit time needs to be considered, and may require extended fecal collections to ensure collection of all unabsorbed isotope.

The precision of the instrument is important for determining how low a level of enrichment can be detected with statistical accuracy. It is common to set the detection limit for isotope ratio increases at three times the standard deviation of the baseline ratio.[3] This is difficult in practice, however, because there is variation among subjects. Fortunately, due to advances in the sensitivity of the instrumentation used to analyze isotope enrichments, especially with ICP-MS, precision is improving and minimal detection limits are decreasing. The magnetic sector TIMS instruments are currently the most sensitive; even the new magnetic sector ICP-MS instruments may never equal the minimal sensitivity of enrichment one can get with the magnetic sector TIMS instruments.

Stable isotopes are obtained as the metal itself, or in the oxide, carbonate, or chloride forms. The isotope needs to be converted into a soluble form prior to administration, and this is usually the soluble chloride or sulphate salt. This is done by dissolving the mineral in a small volume of concentrated acid, either hydrochloric or sulfuric acid. It is important to mention here that all acids used in isotope preparation and sample preparation need to be of ultra-high purity: double sub-boiling, quartz distilled acids are necessary to avoid any trace contaminants. The soluble isotope is diluted to the desired concentration, or further preparation is conducted, if necessary, depending on the mineral, and the form in which it is to be administered. If the isotope is to be administered intravenously, the solutions need to be sterile-filtered, packaged into vials, and tested for sterility and pyrogenicity. This is usually done by a hospital pharmacy experienced in these procedures.

The most common method of isotope administration is using extrinsic labelling. Intrinsic labelling is often done in order to study mineral bioavailability, or to compare the extrinsic label with the intrinsic label to determine possible differences in mineral metabolism. With the extrinsic study approach, the assumption is that the extrinsic mineral tracer exchanges completely with the native mineral, and is absorbed and metabolized identically to the native mineral. Extrinsic labelling is the preferred method to use, due to its simplicity and lower cost. Intrinsic labelling of foods with mineral isotopes is costly, mainly because of low incorporation of the isotope (e.g., 3 to 50%), and the large doses of isotope needed to provide detectable label in the final product. Intrinsic labelling is usually accomplished in plants using hydroponic cultivation, and in animals using oral or parenteral administration. Many human studies have been done comparing intrinsic and extrinsic labelling with mineral isotopes (Zn, Fe, Cu, Se) in a number of food items: poultry,[39] goose,[40] milk,[41,42] beans,[43] and nut butters.[40] Most studies found good agreement between intrinsically and extrinsically administered isotopes of zinc and copper, and recently of non-heme iron used to determine mineral absorption.[3,43] Some studies have reported differences in absorption with the intrinsic and extrinsic tags, in particular with selenium.[44] This difference with Se (absorption of intrinsic label greater than absorption of extrinsic label) may represent differences in the metabolism of the different forms of Se studied (selenite and selenomethionine), since selenite is an anion, and selenomethionine utilizes methionine pathways, representing two separate pools of selenium.[45] This situation may be analogous to heme and non-heme iron pools, and would require labelling of both pools with different isotopes. Based on the results of these studies, very important considerations arise when designing mineral studies utilizing extrinsic labelling. It is crucial that the extrinsic tag has time to fully equilibrate with the native mineral; in addition, the amount of mineral added with the tag will influence the size of the total mineral pool, and may affect its absorption.[3]

## 1.4 Practical Strategies for Conducting Stable-isotope Tracer Studies

### 1.4.1 Zinc

Zinc stable isotopes have been used in human metabolic studies for almost 25 years, with the first few studies published utilizing NAA for isotopic analysis,[3] followed by a rapid increase in the number of zinc studies done due to the development of mass spectrometry techniques for mineral analysis. TIMS and ICP-MS are currently the most common analytical tools, but FAB-MS has been successfully used, and high-resolution ICP-MS is rapidly gaining a foothold.[33] Zinc metabolic studies with stable isotopes have been crucial in understanding zinc absorption, particularly with regard to the ability to distinguish endogenously excreted zinc compared to unabsorbed zinc, which cannot be done using the classic mass balance techniques. Zinc stable isotopes have been used to determine the effects of a multitude of conditions on zinc metabolism: age (premature infants through seniors); dietary factors (bioavailability, supplement use); metabolic state (pregnancy, lactation, fasting, exercise, low zinc status); drug use (oral contraceptives); and disease states (diabetes, liver disease, cystic fibrosis, diarrhea). Zinc stable isotopes have also been used to establish kinetically based compartmental models of zinc metabolism, and a number of these models have been published.[46] These models can serve as useful tools for investigating metabolic systems and processes.

Designing a zinc stable-isotope study is dependent on questions asked concerning zinc. Isotopes can be administered orally and/or intravenously, and there are three possible isotope tracers used: $^{67}Zn$, $^{68}Zn$, and $^{70}Zn$. $^{70}Zn$ is the best tracer to use, as its natural abundance is only 0.62%. Oral tracers alone can be used in studies concerned with bioavailability or other dietary factors such as mineral interactions, and can be useful as estimates of zinc absorption, although there are inherent problems with this technique. The inclusion of an intravenous tracer in combination with an oral tracer vastly improves the information obtained from the study, and is crucial for establishing more detailed compartmental models, determining a better and simpler estimate of zinc absorption, and confirming data obtained with the use of oral tracers. Oral zinc isotopes can be added to the diet either intrinsically or extrinsically, and most studies have shown little difference between the labels when the extrinsic isotope has been allowed to fully exchange with the native element.[3,47]

Sample collection is determined by the type of study being conducted. If the research question involves compartmental modelling, many blood samples are needed following isotope infusion, with other needed biological samples dependent on the model being developed.[38] If the research question involves the determination of fractional zinc absorption only, a minimum of two blood or urine samples is needed if using the dual isotopic tracer ratio method, or complete fecal collections for 7 to 10 days if the fecal monitoring approach is used. It is crucial to avoid any sources of zinc contamination because isotope ratios will be affected.

The determination of fractional zinc absorption is an important analytical measurement, as the homeostatic control over absorption is primarily regulated by the gastrointestinal tract. Two important aspects of zinc homeostasis, exogenous zinc absorption and endogenous zinc secretion, are important for the maintenance of zinc balance, and can adapt to various physiological states.

Therefore, it is important to be able to investigate this quantitatively, which can be accomplished with the use of stable isotopes. There are several stable-isotopic methods used to determine zinc absorption; the recommended method is called the *dual isotopic tracer ratio method*.[48] This method involves the administration of two zinc isotopes; one is given orally (often 67-Zn or 68-Zn), and the other is given intravenously (usually 70-Zn). The minimum sample collection needed is a spot urine sample or a blood sample collected 3 days after isotope administration. This technique was adapted from the established method for calcium absorption, and has been used to determine fractional zinc absorption in a number of studies.[35,48–60] Due to the minimal sample collection (and thus minimal preparation for isotopic analysis) and minimal subject compliance needed, this technique is recommended for determination of fractional zinc absorption, and is especially suitable for studies conducted with a large study population. The classical method of fecal monitoring can also be utilized to determine zinc absorption, particularly when it is prohibitive to use an intravenous isotope, but there are several inherent problems with this method, including sample loss, a high level of required subject compliance, and inability to accurately determine endogenous zinc excretion.

Zinc can be isolated from biological samples (plasma, fecal, milk, urine) following acid digestion and ion exchange chromatography.[38,48] Plasma, fecal, and milk samples can be digested in a microwave system with concentrated nitric acid, which is then evaporated to dryness. Biological samples can also be ashed in a muffle furnace.[61] Digests are then brought up in HCl, and the zinc is isolated using ion-exchange chromatography.[62] Urine samples do not have to be digested in nitric acid; a centrifugation to remove solids, followed by removal of the inorganic salts with a chelating resin, and then ion-exchange chromatography can be used for urine zinc isolation.[48] Zinc from biological samples has also been isolated for isotopic ratio determination using an acid digestion of the material, followed by zinc extraction with a dithiocarbamate.[63] After adjusting the acid digest to pH 4, an ammonium extraction buffer is added, followed by addition of APDC (ammonium pyrrolodin-1-yldithioformate) and $CCl_4$ or DDDC (diethylammonium diethyldithiocarbamate) with tetrachloromethane.[64] The zinc stays with the organic phase, and is extracted again with $CCl_4$ or tetrachloromethane, followed by washing with water and nitric acid and evaporating to dryness several times. Dissolution in nitric acid is the final step for isolation. Another dithiocarbamate that has been used for zinc extraction is ammonium diethyldithiocarbamate.[65]

For determination of zinc isotope ratios using ICP-MS, all pure zinc samples which have been isolated with an HCl matrix are evaporated to dryness and diluted to the same zinc concentration in dilute nitric acid. This concentration depends on the sensitivity of the ICP-MS as well as the sample introduction system. Zinc natural abundance standards, diluted to the same concentration as the enriched samples, are analyzed as standards during the ICP-MS runs, usually every five samples or so, in order to determine natural abundance ratios for correction in the final data analysis. There is a possibility for mass discrimination using the older quadrupole-based ICP-MS instruments for zinc isotope ratios, and this bias can change with time in response to variations in instrumental conditions. Roehl and coworkers used gallium as an isotope ratio internal standard ($^{71}Ga/^{69}Ga$) to correct mass bias drift for ICP-MS Zn isotope ratio determinations.[66] Gallium is suited for mass drift correction in the case of Zn because it is a rare element not present in biological samples, its mass is close to that of Zn and its isotopes are comparable in abundance and do not have isobaric interferences from singly charged ions.

### 1.4.2 Iron

The provision of additional Fe to high risk groups, including infants, toddlers and pregnant women, remains an important means of preventing and treating Fe deficiencies in these groups. Most studies of Fe absorption from either food sources or supplements have been limited by the need to use radioactive Fe to evaluate Fe absorption, or have used indirect measures of Fe metabolism such as hematocrit or serum ferritin to evaluate the effects of Fe supplementation. In radiotracer studies, Fe absorption is determined by measuring fecal recovery, whole-body incorporation, or red blood cell (RBC) incorporation of an orally administered radioactive isotope of Fe (usually $^{59}Fe$).[67,68] Radioisotope techniques are not currently appropriate for use in healthy children, and fecal recovery methods are unlikely to provide reliable results. Recently, we and others have described a two-tracer, stable-isotope method for measuring Fe incorporation. This method is safe for subjects of all ages and allows the accurate comparison of the fraction of Fe which is incorporated into RBCs from two different meals given to the same infant.[2,69–71]

When isotopes are only administered orally, it is impossible to directly assess the actual dietary absorption of iron. Rather, it is necessary to correct the RBC incorporation based on the fraction of isotope which is actually incorporated from the absorbed dose. Usually this fraction is approximately 90%; however, it may be substantially lower in premature infants, pregnant women or patients with anemias related to chronic illness.[71]

One approach to directly assessing this fraction is to give an iron isotope intravenously. However, the calculated absorption may not be accurate if the intravenous isotope is transported differently within the vascular system compared to the absorbed oral isotope. We use Fe-58-citrate as the form of iron to be administered when it is given intravenously. There may be a very small risk of an allergic reaction associated with the use of intravenously administered iron. An adverse reaction is extremely unlikely, due to the form and the small doses of iron used. Because of this potential risk, however, we have chosen to administer iron isotopes slowly by intravenous infusion only within a hospital setting.[2]

Iron samples are prepared for mass spectrometric analysis using ion-exchange techniques. In most studies, iron isotope ratio analysis is performed using TIMS. When three isotopes are used in a study, i.e., Fe-54 and Fe-57 given orally and Fe-58 given intravenously, fractionation correction cannot be applied to the final analysis of the isotope ratios. Fractionation correction is a mathematical adjustment in which the measured isotope ratios are corrected by comparison to the known ratio of two unadministered isotopes.[69] In this case, since there are only four iron isotopes, when three are given in a study, this correction cannot be applied. When the fractionation correction cannot be applied, the accuracy of the final isotope ratios is lowered, depending on the element- and heat-specific magnitude of the fractionation. This is minimized using careful temperature monitoring of the filament mass spectrometer during analysis. However, even using careful temperature control, the accuracy of the non-corrected final enrichment measurement for iron worsens from 0.2% to 1.0% compared with fractionation-corrected measurements.[70]

To allow for this diminished measurement accuracy, when performing a triple isotope experiment, it is necessary to administer doses of isotopes that will result in relatively high enrichment of the tracers. In most cases, it might be preferable to do the studies sequentially; that is, to administer two isotopes at one time, wait at least two weeks and then readminister one or both of these isotopes. Although baseline enrichments of the isotopes would need to be assessed, for most purposes, these do not change substantially over several weeks after reaching a peak 14 days after the initial dosing.[69]

## 1.5 Appendix — Stable-isotope Suppliers

**AMT**, Advanced Materials Technologies, 7a David Devora St., Kiryat Ono 55502, Israel, phone: 972-3-5352039, Fax: 972-3-5344530, *http://www.isotope-amt.com*, e-mail: *sales@isotope.amt.com*.

**Cambridge Isotope Laboratories, Inc.**, 50 Frontage Rd., Andover, MA 01810, U.S.A., phone: 800-322-1174, Fax: 978-749-2768, *http://www.isotope.com*, e-mail: *cilsales@isotope.com*.

**Europa Scientific Ltd.**, Europa House, Electra Way, Crewe CW6 1ZA, U.K., phone: +44 (0) 1270 589398, Fax: +44 (0) 1270 589412.

**C H E M G A S**, 31 bis Avenue Robert Schuman, 92100 Boulogne, France, phone: +33 1 48 25 33 37, Fax: +33 1 48 25 92 40, *http://www.chemgas.com*, e-mail: *chemgas@chemgas.com*.

**JV Isoflex**, 123182, Schukinskaya St. 12-1, Moscow, Russia, phone: 7-095-190 6645, 7-095-158 838, Fax: 7-095-943 0026, *http://www.transit.ru/user/isoflex/*, e-mail: *isotope@isoflex.transit.ru*.

**Novachem Pty., Ltd.**, ACN 005 116 521, 50 Garden St., South Yarra VIC 3141, Australia, *http://www.novachem.com.au*, e-mail: *novachem@novachem.com.au*.

**Oak Ridge National Laboratories**, P.O. Box 2009, Oak Ridge, TN 37831-8044, U.S.A., *http://www.ornl.gov/isotopes/catalog.htm*.

**Pennwood Chemicals, Inc.**, 98 Cuttermill Rd., St. 262, Great Neck, NY, 11021, phone: 516-487-2077, Fax: 516-487-2890, *http://www.pennwoodgroup.com/home.htm*.

**Trace Sciences International Corp.**, 901 Market St., St. 460, Wilmington, DE, U.S.A., phone: 302-426-1590, Fax: 302-429-5953, or 15 Wertheim Ct., St. 404, Richmond Hill, Ontario, L4B 3H7, Canada, phone: 905-707-7000, Fax: 905-707-0700, *http://www.isotopetrace.com*, e-mail: *sales@isotopetrace.com*.

**Urenco Stable Isotopes Business Unit**, Urenco Nederland B.V., P.O. Box 158, 7600 AD ALMELO, The Netherlands, Fax: 31-546-545346, *http://www.urenco.com/isotope/home.htm*, e-mail: *isotopes@urenco.nl*.

## References

1. Clifford, A.J. and Muller, H., Eds. *Mathematical Modelling in Experimental Nutrition*, Plenum Press, New York, NY, 1998.
2. Abrams, S.A., Using stable isotopes to assess mineral absorption and utilization by children, *Am. J. Clin. Nutr.*, 70, 955–964, 1999.
3. Sandström, B., Fairweather-Tait, S., Hurrell, R. et al., Methods for studying mineral and trace element absorption in humans using stable isotopes, *Nutr. Res. Rev.*, 6, 71–95, 1993.
4. Turnlund, J.R., The use of stable isotopes in mineral nutrition research, *J. Nutr.*, 119, 7–14, 1989.
5. Hachey, D.L., Wong, W.W., Boutton, T.W. et al., Isotope ratio measurements in nutrition and biomedical research, *Mass Spectrometry Rev.*, 6, 289–328, 1987.
6. Young, V.R. and Ajami, A., The Rudolf Schoenheimer Centenary Lecture. Isotopes in nutrition research, *Proc. Nutr. Soc.*, 58, 15–32, 1999.
7. Schwartz, R. and Giesecke, C.C., Mass spectrometry of a volatile Mg chelate in the measurement of stable 26 Mg when used as a tracer, *Clin. Chem. Acta*, 97, 1–8, 1979.
8. Buckley, W.T., The use of stable isotopes in studies of mineral metabolism, *Proc. Nutr. Soc.*, 47, 407–416, 1988.
9. Weast, R.C., Ed. *Handbook of Chemistry and Physics*, CRC Press, Inc., Cleveland, OH, 1975.
10. Horlick, G. and Shao, Y., ICP-MS for elemental analysis, in *Inductively Coupled Plasmas in Analytical Atomic Spectrometry*, Montaser, A., Golightly, D.W., Eds., VCH Publishers, Inc., New York, NY, 1992, 551–612.
11. Spitzer, E.J. and Sites, J.R., *Isotopic Mass Spectrometry of the Elements*, Oak Ridge National Laboratories, 1963.
12. De Bievre, P. and Barnes, I.L., Table of the isotopic composition of the elements as determined by mass spectrometry, *Int. J. Mass Spectr. and Ion Proc.*, 65, 211–230, 1985.
13. Abundances, C.o.A.W.a.I., Isotopic composition of the elements 1989, *Pure Appl. Chem.*, 63, 991–1002, 1991.
14. IUPAC, IUPAC Subcommittee on Assessment of Isotopic Composition of the Elements, *Pure Appl. Chem*, 56, 695–768, 1984.
15. Czajka-Narins, D.M., Minerals, in *Food, Nutrition and Diet Therapy,* Mahan, L.K., and Escott-Stump, S., W.B. Saunders Co., Philadelphia, PA, 1996, 123–143.
16. De Laeter, J.R., Heumann, K.G., and Rosman, K.J.R., Isotopic compositions of the elements 1989, *J. Phys. Chem. Ref. Data*, 20, 1327–1337, 1989.

17. Yergey, A.L., Issues in constant tracer infusion and mineral metabolism, in *Nutrient Regulation during Pregnancy, Lactation, and Infant Growth*, Vol. 352; Allen, L., King, J.C., and Lonnerdal, B., Eds., Plenum Press, New York, NY; 1994; 279–290.
18. Stathos, T.H., Shulman, R.J., Schanler, R.J. et al., Effect of carbohydrates on calcium absorption in premature infants, *Pediatr. Res.*, 39, 666–670, 1996.
19. Krebs, N.R., Westcott, J.E., Huffer, J.W. et al., Absorption of exogenous zinc and secretion of endogenous zinc in the human small intestine, *FASEB J.*, 12, A345, 1998.
20. Yergey, A.L., Analytical instruments for stable isotopic tracers in mineral metabolism, *J. Nutr.*, 126, 355S–361S, 1996.
21. Turnlund, J.R., Bioavailability of dietary minerals to humans: the stable isotope approach, *Crit. Rev. Food Sci. Nutr.*, 30, 387–396, 1991.
22. Aggarwal, S.K., Kinter, M., and Herold, D.A., Determination of lead in urine and whole blood by stable isotope dilution gas chromatography: Mass spectrometry, *Clin. Chem.*, 40, 1494–1502, 1994.
23. Aggarwal, S.K., Kinter, M., Fitzgerald, R.L. et al., Mass spectrometry of trace elements in biological samples, *Crit. Rev. Clinical Lab. Sci.*, 31, 35–87, 1994.
24. Aggarwal, S.K., Kinter, M., and Herold, D.A., Mercury determination in blood by gas chromatography-mass spectrometry, *Biol. Trace Elem. Res.*, 41, 89–102, 1994.
25. Montaser, A., McLean, J.A., and Liu, H., An introduction to ICP spectrometries for elemental analysis, in *Inductively Coupled Plasma Mass Spectrometry*, Montaser, A., Ed., Wiley-VCH; New York, NY, 1998, 1–28.
26. Gray, A.L., Plasma sampling mass spectrometry for trace analysis of solutions, *Anal. Chem.*, 47, 600–601, 1975.
27. Janghorbani, M., Ting, B.T., and Fomon, S.J., Erythrocyte incorporation of ingested stable isotope of iron (58Fe), *Am. J. Hematology*, 21, 277–288, 1986.
28. Hamester, M., Wiederin, D., Wills, J., Kerl, W., and C.B. Douthitt, Strategies for isotope ratio measurements with a double focusing sector field ICP-MS, *Fresenius J. Anal. Chem*, 364, 495–497, 1999.
29. Douthitt, C.B., Magnetic sector ICP-MS: Comprehensive bibliographies of HR-ICP-MS and MC-ICP-MS, *ICP Inf. Newsl.*, 25, 87, 1999.
30. Horlick, G. and Montaser, A., Analytical characteristics of ICP-MS, in *Inductively Coupled Plasma Mass Spectrometry* Montaser, A., Ed., Wiley-VCH; New York, NY, 1998, 503–612.
31. Evans, E.H. and Giglio, J.J., Interferences in inductively coupled plasma mass spectrometry, *J. Anal. At. Spectrom.*, 8, 1–18, 1993.
32. Turner, P.J., Mills, D.J., Schroder, E. et al. Instrumentation for low- and high-resolution ICP-MS, in *Inductively Coupled Plasma Mass Spectrometry* Montaser, A., Ed., Wiley-VCH, Inc., New York, NY, 1998, 421–501.
33. Krebs, N.F., Miller, L.V., Naake, V.L. et al., The use of stable isotope techniques to assess zinc metabolism, *J. Nutr. Biochem.*, 6, 292–301, 1995.
34. Hambidge, K.M., Krebs, N.F., and Miller, L., Evaluation of zinc metabolism with use of stable-isotope techniques: implications for the assessment of zinc status, *Am. J. Clin. Nutr.*, 68, 410S–413S, 1998.
35. Abrams, S.A., Wen, J., and Stuff, J.E., Absorption of calcium, zinc, and iron from breast milk by five- to seven-month-old infants [published erratum appears in *Pediatr. Res.* 1997 Jun, 41(6):814], *Pediatr. Res.*, 41, 384–390, 1996.
36. Turnlund, J.R., Keyes, W.R., and Peiffer, G.L., Molybdenum absorption, excretion, and retention studied with stable isotopes in young men at five intakes of dietary molybdenum, *Am. J. Clin. Nutr.*, 62, 790–796, 1995.

37. Turnlund, J.R., Keyes, W.R., Peiffer, G.L. et al., Copper absorption, excretion, and retention by young men consuming low dietary copper determined by using the stable isotope 65Cu, *Am. J. Clin. Nutr.*, 67, 1219–1225, 1998.
38. Lowe, N.M., Shames, D.M., Woodhouse, L.R. et al., A compartmental model of zinc metabolism in healthy women using oral and intravenous stable isotope tracers, *Am. J. Clin. Nutr.*, 65, 1810–1819, 1997.
39. Fairweather-Tait, S.J., Fox, T.E., Wharf, S.G. et al., Zinc absorption in adult men from a chicken sandwich made with white or wholemeal bread, measured by a double-label stable-isotope technique, *Br. J. Nutr.*, 67, 411–419, 1992.
40. Johnson, P.E., Stuart, M.A., Hunt, J.R. et al., $^{65}$Copper absorption by women fed intrinsically and extrinsically labeled goose meat, goose liver, peanut butter and sunflower butter, *J. Nutr.*, 118, 1522–1528, 1988.
41. Serfass, R.E., Lindberg, G.L., Olivares, J.A. et al., Intrinsic labeling of bovine milk with enriched stable isotopes of zinc, *Proc. Soc. Experimen. Biol. Med.*, 186, 113–117, 1987.
42. Serfass, R.E., Ziegler, E.E., Edwards, B.B. et al., Intrinsic and extrinsic stable isotopic zinc absorption by infants from formulas, *J. Nutr.*, 119, 1661–1669, 1989.
43. Donangelo, C.M., Woodhouse, L.R., Mertz, S.D. et al., Both intrinsic and extrinsic iron absorption from a high iron bean variety tends to be lower than from a low iron variety in young women, *FASEB J.*, 13, A242, 1999.
44. Christensen, M.J., Janghorbani, M., Steinke, F.H. et al., Simultaneous determination of absorption of selenium from poultry meat and selenite in young men: application of a triple stable-isotope method, *Br. J. Nutr.*, 50, 43–50, 1983.
45. Aggett, P.J., Iron, copper, and zinc absorption and turnover; the use of stable isotopes, *Eur. J. Pediatr.*, 156 Suppl. 1, S29–34, 1997.
46. Wastney, M.E., Yang, D.C., Andretta, D.F. et al., Distributing working versions of published mathematical models for biological systems via the Internet, *Adv. Experimen. Med. Biol.*, 445, 131–135, 1998.
47. Egan, C.B., Smith, F.G., Houk, R.S. et al., Zinc absorption in women: comparison of intrinsic and extrinsic stable-isotope labels, *Am. J. Clin. Nutr.*, 53, 547–553, 1991.
48. Lowe, N.M., Woodhouse, L.R., Matel, J.S., and King, J.C., Estimation of zinc absorption in humans using four stable isotope tracer methods and compartmental analysis, *Am. J. Clin. Nutr.*, 71, 523–529, 2000.
49. Friel, J.K., Naake, V.L., Jr., Miller, L.V. et al., The analysis of stable isotopes in urine to determine the fractional absorption of zinc, *Am. J. Clin. Nutr.*, 55, 473–477, 1992.
50. English, J.L., Fennessey, P.V., Miller, L.V. et al., Use of a dual isotope technique to measure zinc absorption, *FASEB J.*, 3, A1079, 1989.
51. Morgan, P.N., Woodhouse, L.R., Serfass, R.E. et al., Zinc absorption from a meat-free meal in elderly and younger women using stable isotopes., *FASEB J.*, 7, A279, 1993.
52. Woodhouse, L., Rodrigues, L., Morgan, P. et al., Measurement of zinc fractional absorption with stable isotopes in women fed low or high zinc diets, *FASEB J.*, 6, A1087, 1992.
53. Woodhouse, L.R., Lowe, N.M., Schwandt, J.L. et al., Validation of a dual isotope method to measure zinc absorption, *FASEB J.*, 9, A866, 1995.
54. Schwandt, J.L., Lowe, N.M., Woodhouse, L.R. et al., Variation in zinc absorption in women, *FASEB J.*, 9, A866, 1995.
55. Morgan, P., Woodhouse, L., Abrams, S. et al., Zinc absorption from a meat-based and meatless meal using a dual-isotope method, *FASEB J.*, 8, A918, 1994.

56. Lowe, N.M., Woodhouse, L.R., Schwandt, J.L. et al., Measurement of zinc absorption in humans: a comparison of methods., *FASEB J.*, 9, A866, 1995.
57. Fung, E.B., Ritchie, L.D., Woodhouse, L.R. et al., Iron supplementation inhibits zinc absorption during lactation, *Am. J. Clin. Nutr.*, 61, A113, 1995.
58. Fung, E.B., Ritchie, L.D., Woodhouse, L.R. et al. Zinc metabolism in insulin-dependent diabetic women. in *Trace Elements in Man and Animals — 9: Proceedings of the Ninth International Symposium on Trace Elements in Man and Animals*, Fischer, P.W.F., L'Abbe, M.R., Cockell, K.A., and Gibson, R.A., NRC Research Press, Ottowa, Canada, 1997, 107–109.
59. Friel, J.K., Andrews, W.L., Simmons, B.S. et al., Zinc absorption in premature infants: comparison of two isotopic methods, *Am. J. Clin. Nutr.*, 63, 342–347, 1996.
60. Moser-Veillon, P.B., Patterson, K.Y., and Veillon, C., Zinc absorption in enhanced during lactation, *FASEB J.*, 9, A729, 1995.
61. Sian, L., Mingyan, X., Miller, L.V. et al., Zinc absorption and intestinal losses of endogenous zinc in young Chinese women with marginal zinc intakes, *Am. J. Clin. Nutr.*, 63, 348–353, 1996.
62. Kraus, K.A. and Moore, G.E., Anion Exchange Studies. VI. The divalent transition elements manganese to zinc in hydrochloric acid, *J. Amer. Chem. Soc.*, 75, 1460–1463, 1953.
63. Amarasiriwardena, C.J., Krushevska, A., Foner, H. et al., Sample preparation for inductively coupled plasma mass spectrometric determination of the zinc-70 to zinc-68 isotope ratio in biological samples, *J. Anal. Atomic Spectrom.*, 7, 915–921, 1992.
64. Serfass, R.E., Thompson, J.J., and Houk, R.S., Isotope ratio determinations by inductively coupled plasma/mass spectrometry for zinc bioavailability studies, *Anal. Chim. Acta*, 188, 73–84, 1986.
65. Ramanujam, V.M., Yokoi, K., Egger, N.G. et al., Polyatomics in zinc isotope ratio analysis of plasma samples by inductively coupled plasma-mass spectrometry and applicability of nonextracted samples for zinc kinetics, *Biol. Trace Elem. Res.*, 68, 143–158, 1999.
66. Roehl, R., Gomez, J., and Woodhouse, L.R., Correction of mass bias drift in inductively coupled plasma mass spectrometry measurements of zinc isotope ratios using gallium as an isotope ratio internal standard, *J. Anal. Atom. Spectrom.*, 10, 15–23, 1995.
67. Barrett, J.F., Whittaker, P.G., Fenwick, J.D. et al., Comparison of stable isotopes and radioisotopes in the measurement of iron absorption in healthy women, *Clin. Sci.*, 87, 91–95, 1994.
68. Viteri, F.E. and Kohaut, B.A., Improvement of the Eakins and Brown method for measuring 59Fe and 55Fe in blood and other iron-containing materials by liquid scintillation counting and sample preparation using microwave digestion and ion-exchange column purification of iron, *Anal. Bio.*, 244, 116–123, 1997.
69. Abrams, S.A., Wen, J., O'Brien, K.O. et al., Application of magnetic sector thermal ionization mass spectrometry to studies of erythrocyte iron incorporation in small children, *Biol. Mass Spectrom.*, 23, 771–775, 1994.
70. McDonald, M.C., Abrams, S.A., and Schanler, R.J., Iron absorption and red blood cell incorporation in premature infants fed an iron-fortified infant formula, *Pediatr. Res.*, 44, 507–511, 1998.
71. Abrams, S.A., O'Brien, K.O., Wen, J. et al., Absorption by 1-year-old children of an iron supplement given with cow milk or juice, *Pediatr. Res.*, 39, 171–175, 1996.

# 2

# *Advances in Radioisotope Methodology*

**Marianne Hansen, Mats Isaksson, and Brittmarie Sandström**

**CONTENTS**

2.1 Introduction ... 23
2.2 Radioisotopes ... 24
2.3 Whole-body Counting Techniques ... 28
  2.3.1 Whole-body Counting ... 28
  2.3.2 Whole-body Counting Applications ... 29
    2.3.2.1 Metabolism and Biological Turnover Rate ... 29
    2.3.2.2 Absorption Studies ... 30
  2.3.3 Equipment and Technological Development ... 31
2.4 Body Imaging Techniques ... 34
2.5 Indirect Measurements of Absorption or Metabolism ... 35
  2.5.1 Tissue Retention ... 35
  2.5.2 Urinary Excretion ... 36
  2.5.3 Fecal Monitoring ... 38
  2.5.4 Equipment and Technological Development ... 38
2.6 Conclusion ... 39
References ... 39

## 2.1 Introduction

Radioisotope techniques have been used for analytical, diagnostic, and therapeutic purposes for many decades. Their usefulness in studies of metabolism of essential trace elements in man was also recognized early; some basic nutritional knowledge about iron metabolism originates from radioisotope studies conducted in the1960s.[1–3] Radioisotope techniques have many advantages compared to techniques using non-radioactive tracers. They are true tracer techniques because most radioisotopes can be obtained in almost carrier-free solutions, i.e., labelling of a compound or uptake by tissues will

0-8493-8730-2/00/$0.00+$.50

not change the chemical or metabolic balance. Radioisotopes also allow studies of potentially toxic elements such as mercury and cadmium without increasing the body burden. The detection of radioisotopes is often relatively simple, measurements can be made with a high precision, an analytical blank is seldom needed and pretreatment of samples can often be omitted. For most trace elements, the radioisotopes are cheaper than their non-radioactive counterparts. One of the essential trace elements, manganese, is a mononuclear element, i.e., only one stable isotope is available, and therefore radioisotope techniques are the only alternative for more thorough metabolic studies. Dual or multiple radioisotope techniques can also be used for some elements allowing studies of interactions and of simultaneous intake of the same element in different forms.[4–8] A specific advantage of certain $\gamma$-emitting radioisotopes is the possibility to conduct *in vivo* measurements of body distribution and to follow the metabolism and biological turnover rate in different organs or the total body. These many advantages have to be balanced against the potential hazards of radiation. However, with optimization of measurement conditions and modern equipment, the radiation doses can be kept at levels corresponding to those obtained at common X-ray examinations or long-distance flights.

## 2.2 Radioisotopes

There are some important characteristics of radionuclides, which limit the number of suitable isotopes for human studies. Some of the aspects to consider when choosing a radioisotope, apart from the similarity (e.g., chemical form, biological turnover rate*) with the element to be traced, are

- Physical half-life
- Decay mode
- Photon energy and intensity
- Daughter nuclide
- Availability
- Radiation dose

The physical half-life of radioactive isotopes varies over several orders of magnitude. To be useful in metabolic studies, the half-life of the chosen radioisotope must be sufficiently long to allow for transport of the isotope from the production site to the laboratory and also to make it possible to follow the metabolic process in question. A half-life that is too short could, of course, be compensated for by a higher activity, but this will then increase

* Also called *biological half-life* or the time for half of the administered activity to be excreted from the body.

the radiation dose. For example, $^{47}Ca$ with a half-life of 4.5 days could be used to follow the retention in the human body for about 2–3 weeks. After that, the activity in the body is often too low to permit any accurate measurements and, if the activity is increased to compensate for this, the radiation dose may be unacceptably high.

Radioactive isotopes decay through different processes and the decay mode determines both the method of measurement and the radiation dose. In the α-decay the nucleus emits an α-particle, consisting of two protons and two neutrons, and some of the excess energy is then carried away as kinetic energy by the α-particle. A radioisotope, which decays through α-decay, emits particles with a range in tissue of about 50 μm, depending on the kinetic energy of the particle, and an α-emitting radioisotope thus cannot be detected from outside the body. Also, the use of such radioisotopes inside the human body would cause an unacceptably high radiation dose. Some radioisotopes decay through β-decay, resulting in the emission of an electron ($\beta^-$-particle) or a positron ($\beta^+$-particle) from the nucleus with a range in tissue, which, although larger than the range of α-particles, is insufficient to allow detection outside the body. Uptake of β-emitting radioisotopes, however, may be estimated from measurements of blood, urine, or fecal samples. A third decay process is electron capture (EC), where the nucleus captures one of its orbiting electrons to regain stability.

In connection with α- or β-radiation, radioisotopes often emit γ-radiation and/or X-rays, which can be detected outside the body if the energy of the radiation is sufficient to penetrate the body. Although this radiation is, in fact, electromagnetic radiation (like visible light or radio waves), it can best be described as particles. These particles are called photons and the photon energy is directly proportional to the frequency of the radiation. The nature of light (and other electromagnetic radiation) is thus subject to a wave-particle duality, which is one of the cornerstones in modern physics, dealing with matters on the atomic scale.

The choice of radioisotopes for *in vivo* detection in parts of the body or the whole body thus depends on the ability of the radioisotope to emit γ-radiation of energy suitable for detection outside the body. If the energy is too low, a large amount of the radiation is absorbed within the body and does not reach the detector. A radioisotope may emit γ-radiation of several different energies, which can limit its application if the γ-energies are closely spaced and if the energy-resolution of the detector is insufficient to separate the signals from the different energies. If photons of several energies are emitted following a radioactive decay, each energy has a given probability to be emitted, which is called the intensity. The intensity is often tabulated as the number of photons, with a certain energy, emitted per 100 decays. A low intensity of a photon energy usable for measurement can only be compensated for by increasing the activity and hence the radiation dose.

In some cases, the decay of a radioisotope results in the formation of a radioactive daughter nuclide. This must be taken into account since the daughter nuclide also gives rise to a radiation dose. However, most tables of

**TABLE 2.1**

Some Radionuclides That may be Used in Isotope Studies, Their Half-life, Decay Mode, and Some γ-energies

| Radionuclide | Half-life, $t_{½}$ | Decay Mode | γ-energies (keV) | Daughter Nuclide (decay mode, half-life) | Activity (MBq) for 1 mSv |
|---|---|---|---|---|---|
| $^{28}$Mg | 20.91 h | $β^-$ | 31, 401, 942, 1342 | $^{28}$Al ($β^-$, 2.24 m) | 0.45 |
| $^{41}$Ca | $1.03 \times 10^5$ y | EC | 3.31[a] | $^{41}$K (stable) | 5.26 |
| $^{45}$Ca | 162.61 d | $β^-$ | | $^{45}$Sc (stable) | 1.41 |
| $^{47}$Ca | 4.54 d | $β^-$ | 1297 | $^{47}$Sc ($β^-$, 3.35 d) | 0.62 |
| $^{51}$Cr | 27.70 d | EC | 320 | $^{51}$V (stable) | 26.3[b]<br>27.0[b] |
| $^{52}$Mn | 5.59 d | EC, $β^+$ | 744, 936, 1434 | $^{52}$Cr (stable) | 0.56 |
| $^{54}$Mn | 312.3d | EC $β^+$ (100%)<br>$β^-$ (0.0003%) | 835 | $^{54}$Cr (stable)<br>$^{54}$Fe (stable) | 1.41 |
| $^{55}$Fe | 2.73 y | EC | 5.90[a] | $^{55}$Mn (stable) | 3.03 |
| $^{59}$Fe | 44.50 d | $β^-$ | 1099, 1292 | $^{59}$Co (stable) | 0.56 |
| $^{67}$Cu | 2.58 d | $β^-$ | 185 | $^{67}$Zn (stable) | 2.94 |
| $^{65}$Zn | 244.26 d | EC, $β^+$ | 1116 | $^{65}$Cu (stable) | 0.26 |
| $^{69m}$Zn | 13.76 h | IT, $β^-$ | 439 | $^{69}$Zn ($β^-$, 56.4 m) | 3.03 |
| $^{72}$Zn | 1.94 d | $β^-$ | 145, 192 | $^{72}$Ga ($β^-$, 14.1 h) | 0.71 |
| $^{75}$Se | 119.78 d | EC | 136, 265 | $^{75}$As (stable) | 0.38 |
| $^{115}$Cd | 2.23 d | $β^-$ | 336, 528 | $^{115}$In ($β^-$, $4.41 \times 10^{14}$ y) | 0.71 |
| $^{203}$Hg | 46.61 d | $β^-$ | 279 | $^{203}$Tl (stable) | 0.53[b] (organic)<br>0.91[b] (organic)<br>1.85 (inorganic) |
| $^{203}$Pb | 2.16 d | EC | 279 | $^{203}$Tl (stable) | 4.17 |

*Note:* Also shown is the decay mode and half-life, when appropriate, of the decay products (daughter nuclides) and the activity, which administered orally, will give the radiation dose 1 mSv. In the decay mode column, $^-$ and $^+$ denote negative and positive β-decay, respectively; EC denotes electron capture and IT means internal transition.

Data are taken from References 47 and 48.

[a] X-ray.

[b] Depending on uptake.

dose factors (the radiation dose per unit activity) take into account the dose from subsequent decays. If the daughter nuclide emits γ-radiation of energy close to the mother nuclide, it may be difficult to resolve the two energies in the detector system; this could cause some problems with the quantification.

Table 2.1 presents some commonly used radioisotopes with their physical half-life ($t_{½}$), mode of decay, and most prominent γ-energies. The table also shows if the resulting daughter nuclide is stable or radioactive. Some of the γ-energies given in the table are actually characteristic X-radiation. Other radioisotopes of potential interest in human nutrition research are $^{48}$V ($t_{½}$ 16 d), $^{99}$Mo ($t_{½}$ 2.75 d).

The radiation dose from radioisotopes distributed inside the body depends on a number of factors and is often expressed as the committed effective dose, which is the radiation dose to the whole body received over 50 years. The unit for this radiation dose is the Sievert (Sv). One Sv is a very large dose, and 3 to 4 Sv is a lethal dose for a human if received during a short period of time. Most countries have dose limits for radiological work that are 50 mSv per year or lower. The radiation dose resulting from metabolic studies often lies in the range of a few mSv. Table 2.1 shows the activity that can be administered by oral intake to give a committed effective dose of 1 mSv. For instance, in studies of calcium and zinc absorption in Göteborg, Sweden, 0.2 MBq $^{47}Ca$ and 0.2 MBq $^{65}Zn$ were given together with a meal. The radiation dose from this intake was then about 1.1 mSv. In a second study, 0.1 MBq $^{47}Ca$ was given intravenously, resulting in a radiation dose of about 0.5 mSv. Manganese absorption studies, also performed in Göteborg, gave radiation doses of between 0.1 and 0.8 mSv. As a comparison, the radiation dose in Sweden from natural sources (cosmic radiation, radiation from radioactive elements in the bedrock and from $^{40}K$ in the body) is about 1 mSv/year as an average. If indoor radon and medical treatment and examinations are also included, the mean yearly radiation dose is about 4.5 mSv.

The radiation dose depends on the particular radioisotope, administered activity, physical half-life (as mentioned above), and biological half-life. The radiation exposure to the body of an administered radioisotope is also dependent on the fraction absorbed and, in most tables of dose-factors, a generic uptake is assumed which can vary widely depending on the circumstances in the experiment, e.g., the presence of factors limiting or promoting uptake.

Of almost trivial concern is the availability of the radioisotope. Some radioisotopes are commercially available, but others may have to be produced especially for the investigation. Commercially available isotopes can be purchased from a number of companies specializing in radio pharmaceuticals or from laboratories with access to accelerator or reactor facilities.

In addition to naturally occurring radioisotopes, the development of particle accelerators and nuclear reactors has made it possible to create a number of so-called anthropogenic radioisotopes artificially. The first man-made nuclear reaction, which fulfilled the old dream of the alchemists of transforming one substance into another, was performed by Rutherford in 1919. In this experiment, Rutherford bombarded nitrogen ($^{14}N$) with $\alpha$-particles and ended up with oxygen ($^{17}O$).

Many of the radioisotopes in Table 2.1 are produced by irradiation of a target nuclide by neutrons from a nuclear reactor or a particle accelerator (cyclotron). The target can be a stable, naturally occurring isotope that, in many cases, is enriched due to a low natural abundance, but it can also be another radioactive nuclide. Some of the radioisotopes can also be formed in radioactive decays or as fission fragments in a reactor. For example, $^{47}Ca$ is produced by bombarding $^{46}Ca$ with neutrons in a cyclotron facility, where the neutrons are emitted from a target irradiated with charged particles accelerated in the cyclotron.

## 2.3 Whole-body Counting Techniques

### 2.3.1 Whole-body Counting

Whole-body counting is a method to register and/or quantify radioactive elements within the human body *in vivo*, without sampling or taking biopsies. This demands that the radioactive element in question emit $\gamma$-radiation, since it is not possible to detect particle radiation outside the body. In its simplest form, whole-body counting can be performed by placing a radiation detector close to the body and measuring the signal from the detector. It is, however, desirable to register radiation from a large part of the body; therefore this set-up is often used in certain geometries, such as arch- or chair-geometry. In arch-geometry, the person to be measured is placed on an arch-shaped bed and the detector is placed in or near the center of curvature of the arch. Since arch-geometry can be inconvenient for the person, the measurement is often performed with the individual sitting in a chair instead. Another method to measure a larger part of the body is to use moveable detectors that scan over a person lying on a bed or large detectors that almost completely cover the body. The measurement time is dependent on the equipment used and on the examination and may vary between 1 and 30 minutes.

One important potential source of error in whole-body measurements is the background radiation. Present everywhere, this radiation has its origin in radioactive materials in the ground and in cosmic radiation. Measurements of small amounts of a radioactive substance in the human body therefore require that the background radiation can be reduced as much as possible. This can be achieved with some kind of shielding, which can be constructed to cover the detector and a part of the person or even the whole measurement system, including the person. Many whole-body counters are therefore housed in steel rooms with thick walls of old steel (cast before WWII), and often lined with lead on the inside. The reason for choosing old steel is that new steel contains small amounts of radioactive cobalt, which is used to continuously control the condition of the furnaces in the manufacturing process of the steel. Also, the building material in a whole-body laboratory should be chosen to contain a minimum of radioactive substances. To further reduce the background radiation, the person could take a shower and change clothes before the measurement. This will mitigate the influence on the measurement from decay products of radon on the subject's hair and clothes.

The whole-body counting technique has been used to study absorption and metabolism of trace elements, but also has other applications in which the aim is to register and/or quantify radioactive substances in the body. In the nuclear industry, research facilities, and hospitals, whole-body counting provides a rapid method to check for internal contamination of radioactive substances present at the work place. This type of measurement has also been

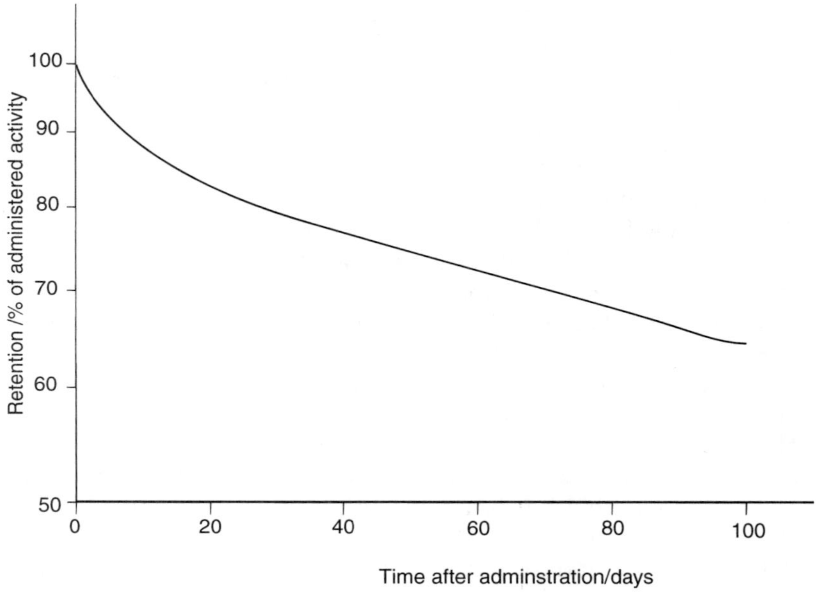


**FIGURE 2.1**
Fractional retention of $^{65}Zn$ at different times after intravenous administration, described by a two-term exponential function.

used extensively in the monitoring of radiation doses from internal contamination after nuclear weapons tests, as well as after the Chernobyl accident.

### 2.3.2 Whole-body Counting Applications

#### *2.3.2.1 Metabolism and Biological Turnover Rate*

The whole-body counting technique is particularly useful for the estimation of metabolism and biological turnover rates of trace elements and minerals. With long-lived radioisotopes such as $^{65}Zn$ and $^{54}Mn$, whole-body retention of a single isotope dose can be measured for up to a year with a reasonable precision.[9,10] Intravenous or oral administration of $^{59}Fe$ and whole-body counting over periods up to 240 days have been used for estimation of the total body iron losses and iron requirements.[1,11,12] In adult men the biological half-life of iron has been found to vary from 500 days to 8.3 years, depending on age.[12,13]

The mean biological turnover rate of zinc in young subjects has been estimated to 247 days from a number of whole-body retention measurements after an i.v. dose of $^{65}Zn$.[14] In a plot of fractional $^{65}Zn$ retention by time, the whole-body turnover rate of zinc was found to follow a two-term exponential function with an initial rapid excretion followed by a slower excretion rate, (Figure 2.1). $^{65}Zn$ measurements in blood and plasma combined with $^{65}Zn$ activity measurements in urine and feces, and, in the whole body after an oral dose, have been used for kinetic studies.[15–17] Through measurements of whole-body $^{65}Zn$ retention and $^{65}$Zn in blood over one year, Watson et al. estimated zinc turnover and zinc content in two body compartments as well as total body zinc.[18]

The biological turnover rate of manganese has been estimated by whole-body counting two to three times weekly for up to 200 days after an oral $^{54}$Mn dose.[10] Manganese turnover was found to fit a single exponential function during the first 10 to 30 days and thereafter a power function resulting in a mean biological half-life of 16 days. When manganese turnover was estimated from whole-body $^{54}$Mn retention measured weekly for 8 weeks, a biological half-life of 30 to 40 days was calculated from the slope of the linear portion of a semi-logarithmic plot of retention vs time.[19]

Selenium turnover has been determined by $^{75}$Se whole-body retention measurements combined with activity measurements in urine and feces up to 40 weeks after an oral dose of [$^{75}$Se]selenomethionine or [$^{75}$Se]selenite.[20,21] Both the whole-body retention and the urine/feces method showed exponential excretion of $^{75}$Se. $^{75}$Se turnover has also been studied by whole-body retention measurements 7 to 22 days after an oral dose and has been found to follow a single exponential function with a mean biological half-life of 30 days in subjects with a habitually low selenium intake.[4]

As with manganese and selenium, copper turnover has been determined as the slope of a semi-logarithmic plot of fractional $^{67}$Cu retention versus time measured over two or three weeks.[22,23] Copper turnover studies with whole-body counting may be useful in diseased individuals; in fact, it has been suggested as a method to identify patients with Wilson's disease, a condition of copper overload. Through several whole-body retention measurements after an intravenous (i.v.) $^{67}$Cu dose, O'Reilly et al. found a markedly prolonged copper turnover in both homo- and heterozygotes for Wilson's disease of 111 and 49 days, respectively, whereas turnover was only 29 days in the control group.[23]

### 2.3.2.2 Absorption Studies

For elements with a long biological half-life, the whole-body retention measurement at a time point when the non-absorbed fraction of a labelled meal or diet has left the body is a close estimate of the degree of absorption. This is the principal method for estimating iron absorption using $^{59}$Fe, often in combination with the β-emitter $^{55}$Fe, which allows dual labelling of two meals or components.[24,25] It is assumed that approximately 80% of absorbed iron is incorporated into red blood cells[26] and this information is utilized in combination with the whole-body measurements to translate the retention of the β-emitting isotope in a blood sample to whole-body retention and thus absorption.

For elements with a more rapid excretion, a correction has to be made for the amount of absorbed isotope which is re-excreted from the time of intake to the time of measurement of the "true" whole-body retention. This is made on the basis of the rate of excretion of an i.v. administered isotope or other estimates of the excretion pattern. When the rate of excretion is reasonably slow, and does not vary considerably between subjects, the mean excretion rate can be used (e.g., for zinc).[14] Other elements (e.g., manganese and copper) show large variations between individuals in turnover rates and thus individual allowances have to be made.[10,27] A single exponential function based on whole-body retention measurements day 10 to 20 or 30 after intake appears applicable for this correction.

When isotopes are used to estimate absorption from foods or diets, a complete isotope exchange between the label and the native element is assumed. Extrinsic labelling has been validated for iron,[28] zinc,[29,30] and manganese.[8] For some other elements, isotope exchange is not likely to occur through extrinsic labelling due to the differences in chemical forms, (e.g., some fortification iron forms and organic selenium forms) and intrinsic (biological) labelling is necessary. When choosing between extrinsic or intrinsic labelling with radioisotopes for absorption studies, the half-life of the isotope has to be encountered. The time-span necessary for biological incorporation of the isotope into a plant or animal will sometimes exclude the use of short-lived radioisotopes. In some cases, this problem may be overcome by application of a higher isotope dose for the labelling, although there is a maximum dose, due to potential radiation damage of the labelled material. Alternatively, a method based on the use of a long-lived isotope may have to be chosen. For example, for the measurement of calcium absorption, $^{45}Ca$ (a β-emitting isotope with $t_{1/2}$ of 162 days), which is measured in a blood sample, may be chosen instead of $^{47}Ca$ ($t_{1/2}$ 4.5 days) measured by whole-body counting.

### 2.3.3 Equipment and Technological Development

In a historical perspective, the use of radium salts in medicine and luminous paint containing radium made it important to find methods to measure the body-burdens of radium, since severe damage was observed among those exposed to it. The first measurements were made by Schlundt et al. in 1929, with a small ionization chamber placed near the subject's spine.[31] Later, in 1937, Evans made measurements with a Geiger-Müller counter, taking the body geometry and background radiation into account.[32] The measurements were performed in an arch-geometry with radius 1 m and the detector placed 1 m from the subject. The minimum detectable amount of radium was further reduced by Hess and McNiff when they started to use a larger ionization chamber.[33] This detector had a volume of 13 liters, which should be compared with the one-liter detector used by Schlundt et al. 18 years earlier. In 1951, when Sievert started to use a circular array of ten long pressurized ionization chambers surrounding the subject, inhomogeneous distribution of a radioisotope in the body was better compensated for and the sensitivity needed to measure γ-radiation from naturally occurring radioisotopes in the body was approached.[34] This equipment was first installed above ground and the shielding against background radiation was done with water tanks. Later, Sievert moved his laboratory below ground level and managed to reduce the influence from the background radiation considerably. In the time period from 1957 to 1960, liquid and plastic scintillators and NaI-detectors began to come into use and further reduced the minimum detectable amount, as well as the measuring time.[35]

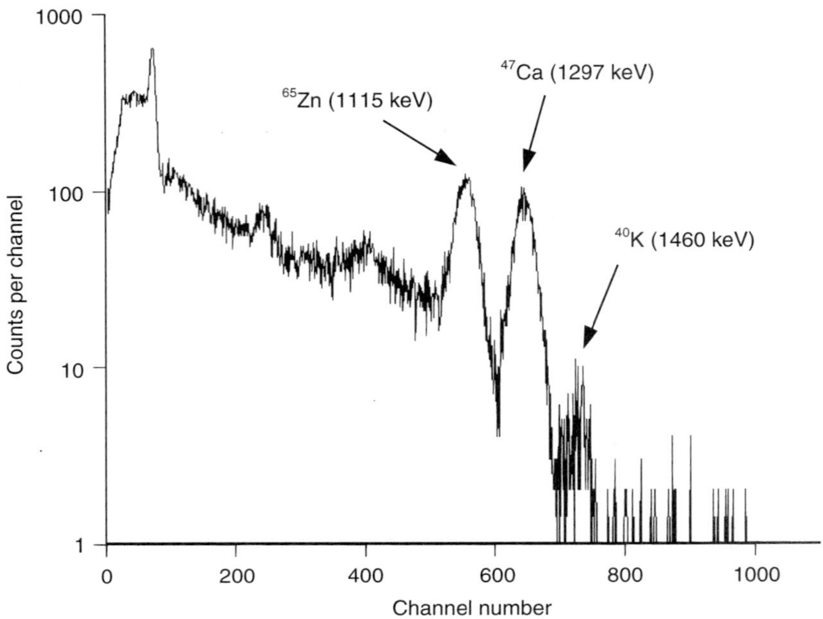


**FIGURE 2.2**
γ-spectrum from a simultaneous measurement of $^{47}Ca$ and $^{65}Zn$ in a whole-body counter.

The development of whole-body counters has proceeded and the detection limit has been reduced through better discrimination against background radiation by housing the subject and detectors in steel chambers and by use of electronic devices. Also, the possibility of making larger NaI-detectors and combinations of different detector materials have improved the sensitivity of the measurements. In addition, the use of large arrays of stationary detectors will increase the sensitivity, but this kind of set-up tends to be very expensive. With the use of NaI and germanium detectors in recent years, the energy resolution has increased, allowing simultaneous measurements on several different radioisotopes with different γ-energies. Two or more radioisotopes may be administered simultaneously if the chosen energy peaks from the different isotopes do not overlap. For example, whole-body retention of $^{54}Mn$, $^{75}Se$ and $^{65}Zn$[4] as well as $^{65}Zn$ and $^{47}Ca$[5–7] has been measured simultaneously (Figure 2.2). Another example is the simultaneous administration of two γ-emitting manganese isotopes $^{52}Mn$ and $^{54}Mn$.[8]

The possible physical set-up for the whole-body counting facilities can be exemplified by the two whole-body counting systems at Sahlgren University Hospital in Göteborg, Sweden. The first system consists of two (ø) 12.7 cm × 10.2 cm NaI(Tl) detectors, mounted in a scanning-bed geometry (Figure 2.3). One detector is mounted above the bed and one below. The detectors and the bed are situated inside a steel room with 150 mm thick walls made of old steel, which is lined with 4 mm lead on the inside. The steel room is also

**FIGURE 2.3**
Photograph showing the whole-body counting facility at Sahlgren University Hospital in Göteborg. The left iron room contains the scanning-bed system with NaI(Tl) detectors and the right iron room houses the plastic scintillators.

supplied with absolute filtered air to avoid variations in background radiation due to airborne radon daughters. The air exchange system also creates an overpressure to prevent accumulation of radon and radon daughters and ensures a constant temperature. To further reduce the background radiation level, the laboratory is built from iron-ore concrete and is situated partly below ground. The detectors are connected to a motor-driven x-y scanning system and during one measurement the detectors may move in a craniocaudal direction for a pre-set time and then, laterally dislocated, in the reverse direction, thus covering the whole person. The total measuring time (usually between 480 and 960 seconds) depends on the experiment. The scanning-scheme and scan-speed can be varied to a great extent.

The second whole-body counting system in Göteborg, also housed in a steel chamber in the same laboratory, comprises four large ($76 \times 92 \times 25$ cm$^3$), plastic scintillation detectors (Figure 2.3). These detectors are stationary, but, because of their size, they detect most of the radiation from the body (about 75%). Due to the poor energy resolution of the plastic detectors, this system is used mainly for determination of potassium by measuring the $\gamma$-radiation from naturally occurring $^{40}$K in the human body. It may, however, also be used for isotope studies with single isotopes.

The choice of detector material depends on the purpose of the investigation and, for isotope research studies, it is often desirable that the detector resolve the different $\gamma$-energies emitted from the radioisotope. In this case NaI(Tl) or some semi-conductor detector is often used. The resolution of a semi-conductor is about ten times the resolution of a NaI-detector, but the sensitivity is often lower due to the smaller size of the detector.

## 2.4 Body Imaging Techniques

A specific feature of radioisotopes is the possibility to visualize the distribution of isotopes in the body and thus follow the turnover rate of trace elements in specific organs and the distribution between organs. When measuring over certain organs or regions of the body, it is important that the detector is properly shielded (collimated) so that the detected radiation originates in the body volume of interest. The design of the collimator depends on the object to be measured and on the γ-energy. If the collimator walls are too thin, radiation that originates outside the region of interest contributes to the signal in an often unforeseeable manner. On the other hand, a lead collimator with thick walls is heavy and can make the experiment difficult to perform. If possible, the collimator opening should cover only the area of interest; however, as the opening is made smaller, the sensitivity decreases, which demands a higher activity or a longer measurement time. In some cases, the blood that passes through the collimator's field of view could carry a certain amount of the radioisotope, which will then interfere with the measurements.

In most whole-body measurements, the electronic equipment is working in a Pulse Height Analysis (PHA) mode. This means that the equipment accepts and processes signals resulting from a wide range of γ-energies and presents the number of registrations for each energy, thus creating a pulse-height spectrum. This spectrum can then be used to identify the contributing radioisotopes from the position of the peaks in the spectrum, as well as the activity from the peak area. For a proper identification and quantification to be made, the system has to be energy- and efficiency-calibrated, using known radioisotopes with well known activities.

Another approach is to collect only those signals generated by γ-radiation of a certain energy and caused by a narrow range of energies during a predetermined time interval — a technique called multichannel scaling (MCS). The pulses are then added during this interval and stored. During a second time interval of the same size, the pulses are again added and stored, etc. With a scanning detector system, the scanning speed can be held constant and the time interval can thus be transformed into a length interval so that, for example, all pulses during each centimeter of the scan are stored in consecutive memory addresses in a computer and presented on a screen. The picture on the screen then shows a profile of the activity distribution from a chosen radioisotope in

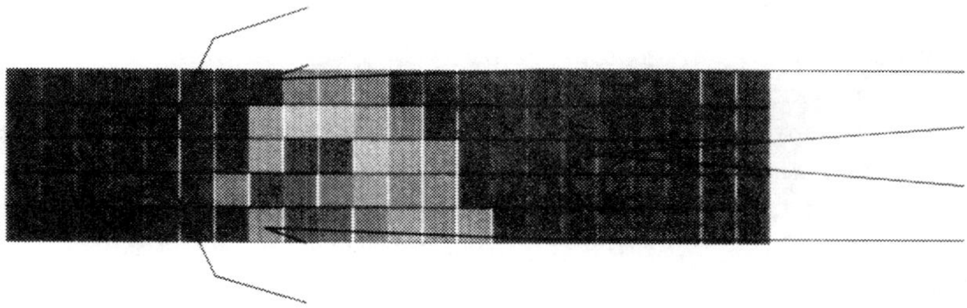

**FIGURE 2.4**
Distribution of $^{54}Mn$ in the body. The figure results from whole-body profile measurements 8 days after oral administration. (With the permission of the *American Journal of Clinical Nutrition*.)

the body. With the proper electronic equipment, it is possible to make profiles of several radioisotopes in the body simultaneously.[36]

The method above presents a one-dimensional profile of the activity distribution, but if several scans are made, with each scan transversally dislocated from the other, a two-dimensional map can be made (Figure 2.4).[10] The detector has to be equipped with a proper collimator to prevent oblique $\gamma$-rays from contributing to the signal. Since the penetrating power of the $\gamma$-radiation depends on the energy, the collimator thickness has to be optimized for each radioisotope studied. The collimator opening limits the sensitivity and thus the scanning speed has to be optimized for each measurement situation.

## 2.5 Indirect Measurements of Absorption or Metabolism

### 2.5.1 Tissue Retention

Appearance of orally administered radioisotopes in plasma or blood has been used to estimate the relative absorption of trace elements. Iron absorption can be estimated from retention of radioisotopes in red blood cells using an estimate or separate measurement of the blood volume and the assumption that 80% of the absorbed iron is incorporated into hemoglobin.[26,28] Zinc absorption has been estimated by a deconvolution method based on measurement of plasma appearance of $^{69m}Zn$ after an oral dose and plasma disappearance of $^{69m}Zn$ after an i.v. dose in the same subjects, when the two measurements are separated in time by two weeks.[37]

Change in radioactivity in plasma or blood over time is also a proximate for the whole-body turnover. Iron metabolism has been followed by measurements of $^{59}Fe$ activity in blood samples over several years, and zinc metabolism has been studied in a similar way combined with measurements of urinary and fecal excretion.[13,15–17]

### 2.5.2 Urinary Excretion

Some minerals are excreted to a relatively large extent in urine. This may be exploited in absorption studies; however, systematic investigations of the relationship between urinary radioisotope excretion and absorption are lacking.

In single-meal studies, urinary $^{47}Ca$ activity measured in a well-type $\gamma$-counter, has been found to correlate significantly with $^{47}Ca$ absorption measured by whole-body counting (Figure 2.5).[38] Typically 2 to 4% of the administered $^{47}Ca$ dose is excreted in urine during the first 48 hours. This may be a fast, precise, and also cheap method for measurement of relative calcium absorption when a whole-body counter is not available. It is also likely that relative absorption of magnesium could be estimated through $^{28}Mg$ excretion in urine, due to the large urinary excretion of magnesium. This method could be particularly relevant for magnesium as whole-body counting is limited by the fast decay of $^{28}Mg$ ($t_{1/2}$ 20.9 hours).

Selenium is also excreted to a large extent in urine. $^{75}Se$ activity of urine collected during two weeks after an isotope dose has been found to be twice as high after administration of selenite (14 to 20% of the dose) compared to when selenomethionine was given (6 to 9% of the dose).[20–21] Thus urinary selenium isotope excretion may be a good indicator of the relative bioavailability of different selenium forms. Combined measurements of $^{75}Se$ in urine and plasma over time in subjects given an oral $^{75}Se$ dose may provide information about selenium metabolism. A fast appearance of $^{75}Se$ in urine and a continuing rise in plasma activity during the first 7 to 12 hours after oral administration suggest accumulation of absorbed selenium in a functional compartment of blood.[21] Although typically less than 0.5% of an administered dose of $^{65}Zn$ is excreted in urine within the first few days after an oral dose, $^{65}Zn$ is excreted in urine proportionally to $^{65}Zn$ absorption found by the whole-body counting method (Figure 2.5).[38]

The combination of whole-body retention measurements with the specific activity of $^{75}Se$ in urine has been used to estimate total body content of selenium.[39] In a similar way, combination of whole-body measurements with measurements on urine samples has been used to investigate the relationship between whole-body content and urine activity of radioactive $^{137}Cs$. The whole-body activity of $^{137}Cs$ could be estimated from the activity in urine samples to provide a good alternative in situations where a whole-body counter is not available.[40]

The use of a very long-lived radioisotope, $^{41}Ca$ ($t_{1/2}$ 103,000 y) has been introduced in a novel method for measurement of bone turnover and calcium metabolism. The bone calcium is labelled through an i.v. administration of the radionuclide. After a certain equilibration time, $^{41}Ca$ content in urine and blood may be followed during, for example, an intervention over as long time period as the study requires.[41] This may be the first realistic method by which calcium metabolism can be followed over a lifetime.

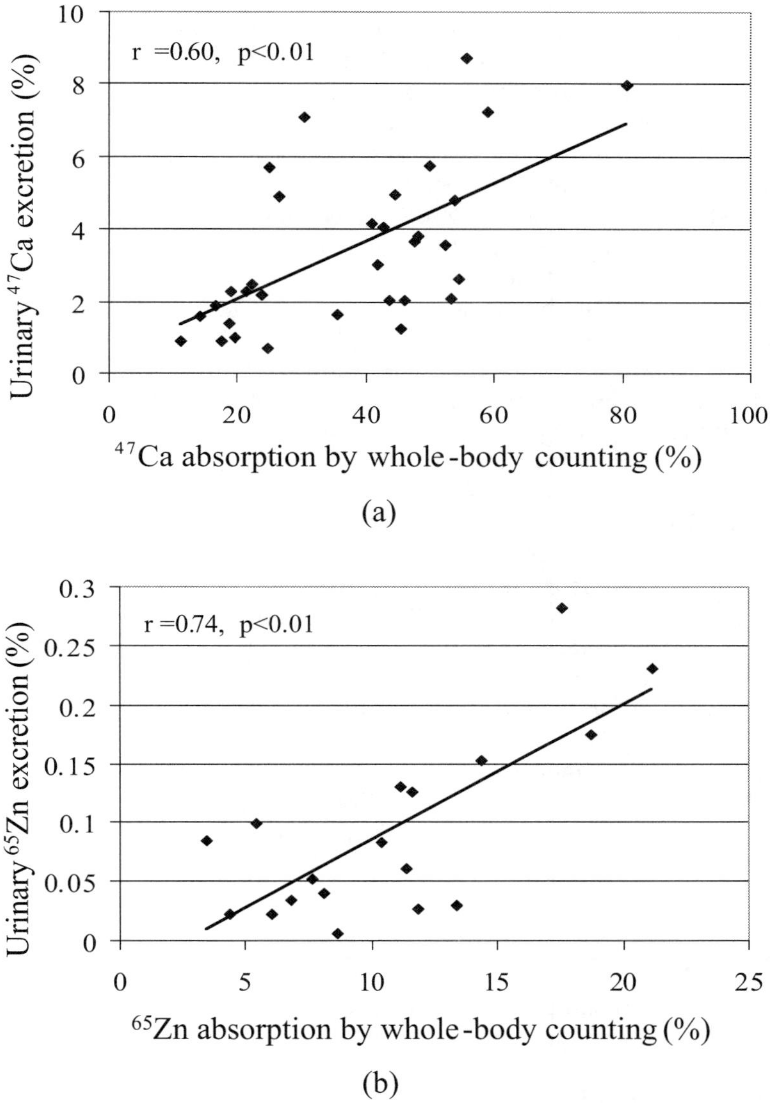


**FIGURE 2.5**
Correlation between: (a) urinary excretion of $^{47}$Ca during 48 hours after an oral $^{47}$Ca dose and absorption measured by whole-body counting, n = 32, and (b) urinary $^{65}$Zn excretion during 48 hours and $^{65}$Zn absorption measured by whole-body counting, n = 18.

### 2.5.3 Fecal Monitoring

Apparent trace element absorption from an oral radiolabelled dose can be determined by measurements of activity in fecal collections. This fecal monitoring technique may, in principle, be used for all trace elements; however, it is less precise for trace elements with a low absorption (e.g., iron and manganese) compared to those with a higher absorption (e.g., zinc and copper). The method has been validated against the whole-body counting method for zinc by Knudsen et al. by the use of $^{65}Zn$.[42] Corrections for re-excretion of absorbed $^{65}Zn$ can be made by a compartmental model developed from $^{65}Zn$ kinetic studies by Wastney and Henkin.[43] This model allows adjustment for excretion of tracers taken up by intestinal cells and later excreted without entering the blood circulation. With a combination of whole-body retention measurements and fecal radioisotope excretion, additional information about metabolism under different dietary conditions can be obtained.[44]

Fecal monitoring has also been used to study copper metabolism in disease. Strickland et al. found a lower 5-day $^{67}Cu$ fecal excretion after an i.v. dose in patients with Wilson's disease compared to control subjects.[27] This suggested decreased biliary copper excretion as part of the explanation for accumulation of copper in these patients.

The non-absorbable radioisotope $^{51}Cr$ may be applied as a stool marker. The fecal activity, measured in a well-type $\gamma$-counter, has been found to follow excretion of simultaneously administered radio-opaque pellets.[10] Thus, $^{51}Cr$ may be used to demonstrate complete intestinal excretion of unabsorbed isotope in fecal monitoring studies and in whole-body counting studies in order to estimate absorption after an oral dose.[10,30] The ratio between $^{51}Cr$ and $^{65}Zn$ in a single stool specimen 24 to 72 hours after intake of the isotopes has been found to be closely correlated to whole-body counting 7 days later and might be a simple method for zinc absorption measurements when a whole-body counter is not available.[45]

### 2.5.4 Equipment and Technological Development

Tissue and excreta samples can be measured with different kinds of equipment. For example, $\gamma$-emitting radionuclides could be analyzed with a NaI-detector or a Ge-detector that was placed in a lead cave to reduce the influence from background radiation. The samples can be measured in different geometries, depending on the sample activity. It is important that the detector system is calibrated for each geometry used. Urine samples may, for example, be measured in 5-liter plastic cans and placed in a holder close to the detector, while smaller samples may be analyzed using a well-type NaI-detector.

$\beta$-emitting radioisotopes, as well as radioisotopes emitting X-rays, may also be measured in tissue samples as long as the samples are treated to allow for the weakly penetrating $\beta$-particles and X-rays to reach the detector. The samples may then be evaporated, ashed, or chemically dissolved to reduce self-absorption in the sample; the radioisotope can even be chemically extracted. The measurements are often made in liquid scintillation counters where the sample is mixed with the detector material. The β-particles thus will not have to penetrate any kind of detector cover.

Radioisotopes with very long half-lives, e.g., $^{41}Ca$, may be difficult to detect by radiation measurements since the long half-lives make the radioactive decays very sparse. An alternative to counting decaying radionuclides is to count the atoms in a sample with the aid of a particle accelerator. This method is called accelerator mass spectrometry (AMS) and has been used for $^{14}C$ in studies of fat metabolism as well as dating of ancient objects.[46] The method has a very low detection limit, implying that the radiation dose can be low and that only small samples are needed.

## 2.6 Conclusion

The advances in radioisotope methodologies are, to a large extent, related to the development of new detectors with high sensitivity and ability to discriminate between isotopes, as well as sophisticated software for the analyses and presentation of the measurements. Even with less expensive equipment, however, reliable information about the metabolic handling of trace elements can be obtained from measurements of blood concentrations or urinary and fecal excretion. For many of the essential trace elements, the validity and full potential of these more indirect techniques require further methodological studies with simultaneous whole-body measurements. Other areas which deserve to be further explored include use of radioisotopes with long physical half-lives, advanced mass-spectrometric techniques for detection, and combined use of radioisotopes and stable isotopes.

## References


1. Bothwell, T. and Finch, C.A., Iron losses in man, in *Occurrence, causes and prevention of nutritional anaemias. Symposia of the Swedish Nutrition Foundation, VI.*, Blix G., Ed., Almquist & Wiksell, Stockholm, 1968, 104.
2. Hallberg, L. and Rossander-Hultén, L., Iron requirements in menstruating women, *Am. J. Clin. Nutr.*, 54, 1047, 1991.
3. Green, R. et al., Body iron excretion in man, *Am. J. Med.*, 45, 336, 1968.
4. Sandström, B. et al., Retention of selenium ($^{75}Se$), zinc ($^{65}Zn$) and manganese ($^{54}Mn$) in humans after intake of a labelled vitamin and mineral supplement, *J. Trace Elem. Electrolytes Health Dis.*, 1, 33, 1987.
5. Hansen, M. et al., Effect of casein phosphopeptides on zinc and calcium absorption from bread meals, *J. Trace Elem. Med. Biol.*, 11, 143, 1997.
6. Sandström, B. et al., Retention of zinc and calcium from the human colon, *Am. J. Clin. Nutr.*, 44, 501, 1986.

7. Hansen, M. et al., Casein phosphopeptides improve zinc and calcium absorption from rice-based but not from whole-grain infant cereal, *J. Pediatr. Gastroenterol. Nutr.*, 24, 56, 1997.
8. Davidsson, L. et al., Intrinsic and extrinsic labeling for studies of manganese absorption in humans, *J. Nutr.*, 118, 1517, 1988.
9. Lykken, G.I., A whole body counting technique using ultralow doses of $^{59}$Fe and $^{65}$Zn in absorption and retention studies in humans, *Am. J. Clin. Nutr.*, 37, 652, 1983.
10. Davidsson, L. et al., Manganese retention in man: A method for estimating manganese absorption in man, *Am. J. Clin. Nutr.*, 49, 170, 1989.
11. Saito, H. et al., Whole-body iron loss in normal man measured with a gamma spectrometer, *J. Nucl. Med.*, 5, 571, 1964.
12. Bonnet, J.D. et al., Rate of loss of radioiron from mouse and man, *Am. J. Physiol.*, 198, 784, 1960.
13. Finch, C.A., Body iron exchange in man, *J. Clin. Invest.*, 38, 392, 1959.
14. Arvidsson, B. et al., A radionuclide technique for studies of zinc absorption in man, *Int. J. Nucl. Med. Biol.*, 5, 104, 1978.
15. Wastney, M.E. et al., Kinetic analysis of zinc metabolism in humans after simultaneous administration of $^{65}$Zn and $^{70}$Zn, *Am. J. Physiol.*, 260, R134, 1991.
16. Babcock, A.K. et al., Effects of oral zinc loading on zinc metabolism in humans II: *In vivo* kinetics, *Metabolism*, 31, 335, 1982.
17. Wastney, M.E. et al., Kinetic analysis of zinc metabolism and its regulation in normal humans, *Am. J. Physiol.*, 251, R398, 1986.
18. Watson, W.S. et al., A two-compartment model for zinc in humans, *J. Trace Elem. Med. Biol.*, 13, 141, 1999.
19. Johnson, P.E., Lykken, G.I., and Korynta, E.D., Absorption and biological half-life in humans of intrinsic and extrinsic $^{54}$Mn tracers from foods of plant origin, *J. Nutr.*, 121, 711, 1991.
20. Griffiths, N.M., Stewart, R.D.H., and Robinson, M.F., The metabolism of [$^{75}$Se]selenomethionine in four women, *Br. J. Nutr.*, 35, 373, 1976.
21. Thomson, C.D. and Stewart, R.D.H., The metabolism of [$^{75}$Se]selenite in young women, *Br J. Nutr.*, 32, 47, 1974.
22. Johnson, P., Milne, D.B., and Lykken, G.I., Effects of age and sex on copper absorption, biological half-life, and status in humans, *Am. J. Clin. Nutr.*, 56, 917, 1992.
23. O'Reilly, S. et al., Abnormalities of the physiology of copper in Wilson's disease, *Arch. Neurol.*, 24, 385, 1971.
24. Bukhave, K., Sørensen, A.D., and Hansen, M., A simplified method for determination of radioactive iron in whole-blood samples, *J. Trace Elem. Med. Biol.*, in press.
25. Eakins, J.D. and Brown, D.A., An improved method for the simultaneous determination of $^{55}$Fe and $^{59}$Fe in blood by liquid scintillation counting, *Int. J. Appl. Radiat. Isot.*, 17, 391, 1966.
26. Hosain, F., Marsaglia, G., and Finch, C.A., Blood ferrokinetics in normal man, *J. Clin. Invest.*, 46, 1, 1967.
27. Strickland, G.T. et al., Turnover studies of copper in homozygotes and heterozygotes for Wilson's disease and controls: Isotope tracer studies with $^{67}$Cu, *Clin. Sci.*, 43, 605, 1972.
28. Hallberg L., Food Iron Absorption, in *Methods in Hematology*, Cook J.D., Ed., Churchill-Livingstone, New York, 1980, chap. 6.

29. Gallaher, D.D. et al., Bioavailability in humans of zinc from beef: Intrinsic vs. extrinsic labels, *Am. J. Clin. Nutr.*, 48, 350, 1988.
30. Flanagan, P.R. et al., Dual-isotope method for determination of human zinc absorption: The use of a test meal of turkey meat, *J. Nutr.*, 115, 111, 1985.
31. Schlundt, H., Barker, H.H., and Flinn, F.B., The detection and estimation of radium and mesothorium in living persons, *Am. J. Roentgenol.*, 21, 345, 1929.
32. Evans, R.D., Radium poisoning. II. The quantitative determination of the radium content and radium elimination rate of living persons, *Am. J. Roentgenol.*, 37, 368, 1937.
33. Hess, V.F. and McNiff, W.T., Quantitative determination of the radium content of the human body and of the radon content of breath samples for the prevention and control of radium poisoning in persons employed in the radium industry, *Am. J. Roentgenol.* 57, 91, 1947.
34. Sievert, R.M., Measurements of $\gamma$-radiation from the human body, *Ark. Fys.* 3, 337, 1951.
35. Spiers, F.W., Whole-body counting: an introductory review, in *Proc. Symp. Whole-body Counting*, International Atomic Energy Agency, (IAEA), Vienna, 1962, 3.
36. Isaksson, M. et al., *In vivo* identification and localisation of radioactive contamination in the human body, in *Radiat. Prot. Dosim.*, 89, 317, 2000.
37. Molokhia, M. et al., A simple method for measuring zinc absorption in man using a short-lived isotope ($^{69m}Zn$), *Am. J. Clin. Nutr.*, 33, 881, 1980.
38. Hansen M. et al., unpublished data, 2000.
39. Stewart, R.D.H. et al., Quantitative selenium metabolism in normal New Zealand women, *Br. J. Nutr.*, 40, 45, 1978.
40. Wallström, E., *Assessment of Population Radiation Exposure after a Nuclear Reactor Accident*, Ph.D. Thesis, Göteborg University, Sweden, 1998.
41. Johnson, R.R. et al., Calcium resorption from bone in a human studied by $^{41}Ca$ tracing, *Nucl. Instr. Meth. Phys. Res.*, B92, 483, 1994.
42. Knudsen, E. et al., Zinc absorption estimated by fecal monitoring of zinc stable isotopes validated by comparison with whole-body retention of zinc radioisotopes in humans, *J. Nutr.*, 125, 1247, 1995.
43. Wastney, M.E. and Henkin, R.I., Calculation of zinc absorption in humans using tracers fecal monitoring and a compartmental approach, *J. Nutr.*, 119, 1438, 1989.
44. Sandström, B., Madsen, E., and Cederblad, Å., Rate of endogenous zinc excretion at high phytate intake, in *Trace Elements in Man and Animals — TEMA 8*, Anke, M., Meissner, D., and Mills, C.F., Eds., 1993, 620.
45. Payton, K.B. et al., Technique for determination of human zinc absorption from measurement of radioactivity in a fecal sample or the body, *Gastroenterology*, 83, 1264, 1982.
46. Stenström, K. et al., Application of Accelerator Mass Spectrometry (AMS) for high-sensitivity measurements of $^{14}CO_2$ in long-term studies of fat metabolism, *Appl. Radiat. Isot.*, 47, 417, 1996.
47. Chu, S.Y.F., Ekström, L.P., and Firestone, R.B., WWW Table of Radioactive Isotopes, database version 2/28/99, from http://nucleardata.nuclear.lu.se/nucleardata/toi/, 1999.
48. International Commission on Radiological Protection, *Age-dependent Doses to Members of the Public from Intake of Radionuclides: Part 5 Compilation of Ingestion and Inhalation Dose Coefficients*, ICRP Publication 72, Pergamon Press, Oxford, U.K., 1996.

# 3

# *Tracer-to-tracee Ratio for Compartmental Modelling of Stable-isotope Tracer Data*

**Gianna Toffolo, David M. Shames, Alessandro Stevanato, and Claudio Cobelli**

**CONTENTS**

3.1 Introduction ........ 43
3.2 Single-pool Tracer Kinetics and Measurement ........ 44
3.3 Tracer-to-tracee Ratio from Mass Spectrometry Measurements ........ 47
3.4 Multi-pool Tracer Kinetics and Measurement ........ 50
3.5 The Multiple Tracer Case ........ 52
3.6 A Test of the Endogenous-constant, Steady-state Assumption ........ 54
3.7 Software Tool: TTRM ........ 54
3.8 Conclusion ........ 56
References ........ 56

## 3.1 Introduction

Compartmental modelling of tracer data is an important tool to provide an *in vivo* measure of nonaccessible parameters and variables, e.g., masses and fluxes.[1,2] In mineral kinetic studies, they have allowed a detailed quantification of the metabolism of many mineral elements in humans. Traditionally, radioactive tracers were used to this purpose.[3–5] More recently, the use of stable isotopes as an alternative to radioactive isotopes is gaining in popularity, due to an increasing awareness of their advantages in biomedical research. In principle, a stable-isotope approach is possible for most studies, since almost all of the minerals relevant to human nutrition have more than a single stable isotope.[6] At present, stable-isotope tracer experiments have already provided the basis for compartmental models of a number of mineral studies, including zinc, selenium, and copper.[7–10]

0-8493-8730-2/00/$0.00+$.50

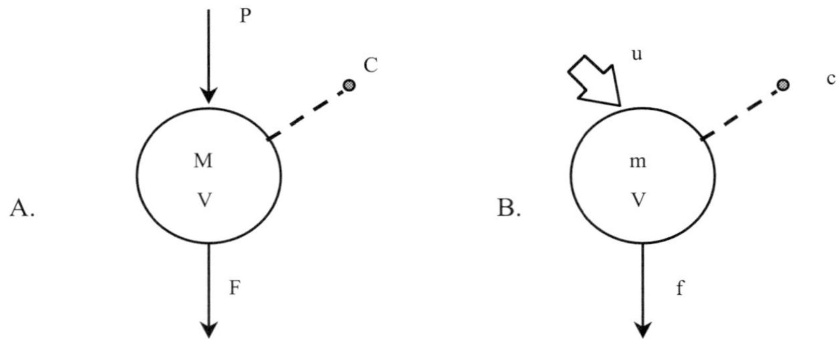


**FIGURE 3.1**
(A) The single compartment tracee system. (B) The single compartment tracer system.

By using some results on isotopic tracers, we have recently proposed a compartmental analysis of stable isotope zinc data by using the tracer-to-tracee ratio (ttr) as the measurement variable, also as suggested by others.[8,11–13] The purpose here is to show that this variable establishes the link with the formalism available for radioactive tracer data, when the endogenous system is in constant steady state. To do that, we first discuss some fundamentals of compartmental analysis of tracer kinetics in general, and then in relation to radioactive and stable-isotopic tracers, with particular attention to measurement. Simple conversion procedures, implemented algorithmically in a software package, are presented to derive in practice the tracer-to-tracee ratio from mass spectrometry measurements. For sake of simplicity, we consider first a single pool system probed with a single tracer. Next, the extension to multipool systems and/or to multiple tracer experiments is outlined. Finally, we propose a test of the constant steady state assumption of the endogenous system when a non-negligible mass stable-isotope tracer is used.

## 3.2 Single-pool Tracer Kinetics and Measurement

Define as "tracee" the endogenous material naturally present in the system, and as "tracer" the infused material. Consider an experiment consisting of a tracer input (Figure 3.1) administered to a single-pool tracee system in a constant steady state. The tracer mass is not assumed to be negligible with respect to the tracee mass, but we assume that it does not perturb the tracee steady state and that it is metabolically indistinguishable from the tracee.

Because of the constant steady state, tracee mass M, production P (mass/time), and disposal F (mass/time) are constant and the mass balance equation for the tracee is

$$\frac{dM}{dt} = P - F = 0 \tag{3.1}$$

By denoting with u the tracer input, m its mass and f its disposal, the mass balance equation for the tracer is

$$\frac{dm(t)}{dt} = u(t) - f(t) \tag{3.2}$$

The link between the tracer and tracee is dictated by the tracer-tracee indistinguishability principle, i.e., the probability that the tracer leaves the pool is equal to the probability that a particle in the pool is a tracer. This can be written as

$$\frac{f(t)}{F + f(t)} = \frac{m(t)}{M + m(t)} \tag{3.3}$$

from which one obtains

$$f(t) = \frac{F}{M}m(t) \tag{3.4}$$

When this expression for f is substituted into Equation 3.2, the tracer model becomes

$$\frac{dm(t)}{dt} = u(t) - f(t) = u(t) - \frac{F}{M}m(t) = u(t) - k \cdot m(t) \tag{3.5}$$

where k = F/M is the rate constant ($time^{-1}$). This equation is a linear, constant coefficient differential equation which provides the link between the tracer and tracee systems since the tracer parameter k reflects tracee events. In order to derive a numerical estimate of k from tracer data, an equation must be associated to Equation 3.5 to relate the tracer mass to the measurement variable, since neither for radioactive nor for stable-isotope tracers is the measured variable the tracer mass.

Consider a radioactive tracer. In this case, measurement usually is tracer concentration, quantitated in terms of energy (dpm) per unit volume; therefore, the measurement equation has the form

$$c(t) = \frac{m(t)}{V} \tag{3.6}$$

where V is the volume of distribution of the pool, which, due to indistinguishability of tracer and tracee, is the same for both.

By using Equations 3.5 and 3.6, the measurement variable c can be expressed as a function of the unknown parameters. For instance, if the tracer experiment consists of injecting the radioactive tracer as a bolus of dose d at time zero, then the solution of Equation 3.5 is

$$m(t) = d \cdot e^{-kt} \tag{3.7}$$

Using Equation 3.7 in Equation 3.6, the expression for tracer concentration c is

$$c(t) = \frac{m(t)}{V} = \frac{d}{V} \cdot e^{-kt} \tag{3.8}$$

where the unknown parameters are the volume V and the exponential k. Both parameters can be estimated from the tracer data: the ratio d/V equals the tracer concentration at time zero, whence

$$V = \frac{d}{c(0)} \tag{3.9}$$

while k can be estimated from the rate of plasma disappearance of the tracer. From the estimates of k and V, it is straightforward to calculate the tracee mass ($M = C \cdot V$) and fluxes ($P = F = kM$) knowing the tracee concentration C. This latter variable coincides in practice with the concentration $C_{tot}$ measured during the experiment, since the radioactive tracer is usually administered in negligible amounts compared with the tracee.

The situation with stable-isotope tracers is different from the radioactive case since (a) stable-isotope tracers are usually introduced into a system in non-negligible amounts, (b) both tracer and tracee consist of mixtures of the same isotopes at different proportion, and (c) mass spectrometry techniques quantify mass ratios between different isotopes, from which different variables such as isotope ratios, abundances, enrichment, and tracer-to-tracee ratio can be derived. Among these variables, tracer-to-tracee ratio, denoted as ttr, is the most convenient way to express the stable-isotope tracer measurement

$$ttr(t) = \frac{m(t)}{M} \tag{3.10}$$

because Equation 3.10 is similar to the radioactive counterpart, Equation 3.6, with M in place of V. Therefore, the use of ttr allows one to write the stable isotope tracer model equations (Equations 3.5 and 3.10) with a formalism similar to the one in use with radioactive tracer (Equations 3.5 and 3.6). It follows that parameters k and M can be estimated for the radioactive tracer, since the solution for ttr is very similar to the solution for c:

$$ttr(t) = \frac{m(t)}{M} = \frac{d}{M}e^{-kt} \tag{3.11}$$

From k and M, tracee fluxes can be calculated as $P = F = kM$.

The analysis above has allowed us to show that a similar formalism can be adopted for radioactive or stable-isotope tracers to estimate the turnover rate of a substance if the tracer measurement is expressed as tracer concentration c or tracer-to-tracee ratio (ttr), respectively. It is of interest to note that the ttr formalism also applies to radioactive tracers, since ttr is equivalent to specific activity, a variable frequently used to express radioactive tracer data. In fact, specific activity, denoted as sa, is defined as the ratio of m to m+M and virtually coincides with ttr since radioactive tracer mass m is usually negligible with respect to tracee mass M:

$$sa(t) = \frac{m(t)}{M + m(t)} \approx \frac{m(t)}{M} = ttr(t) \tag{3.12}$$

From Equation 3.12, sa and thus ttr can be evaluated as the ratio of two primary measurements: tracer and tracee concentrations. For stable-isotope tracers, the relationships between ttr and primary mass spectrometry measurements is less straightforward and will be discussed in the following section.

## 3.3 Tracer-to-tracee Ratio from Mass Spectrometry Measurements

Consider an element with at least two stable isotopes, and denote with superscript a the isotope used as a reference, e.g., the most naturally abundant isotope or the isotope having the lower mass. Other isotopes will be denoted by a superscript equal to the difference, in mass units, between their mass and that of the reference isotope. For example, consider Zn which has five isotopes: $^{64}Zn$, $^{66}Zn$, $^{67}Zn$, $^{68}Zn$, and $^{70}Zn$: if $^{66}Zn$ is chosen as a reference, the other isotopes are denoted by superscript $^{(-2), (1), (2)}$, and $^{(4)}$, respectively.

All the stable isotopes are present in the tracee and tracer: in the tracee, they are at natural proportions; in the tracer, one isotope is artificially enriched above its natural level. Tracee and tracer mass can be subdivided into their isotopic components, e.g., for the Zn tracers: $M = M^{(a)} + M^{(-2)} + M^{(1)} + M^{(2)} + M^{(4)}$; $m = m^{(a)} + m^{(-2)} + m^{(1)} + m^{(2)} + m^{(4)}$. Finally, denote by superscript k the isotope which is artificially elevated in the tracer, e.g., k = 1 for a $^{67}Zn$ tracer, k = 2 for a $^{68}Zn$ tracer.

Based on the above definitions, the isotope ratios are the quotient of the amount of each isotope and isotope $^{(a)}$, i.e., for isotope $^{(k)}$, the isotope ratio in a sample taken during the tracer experiment, when both tracee and tracer are simultaneously present, is

$$r^{(k)}(t) = \frac{M^{(k)} + m^{(k)}(t)}{M^{(a)} + m^{(a)}(t)} \tag{3.13}$$

The isotope abundances are the quotient of the amount of each isotope and the total mass (sum of all isotopes), i.e., for isotope $^{(k)}$

$$a^{(k)}(t) = \frac{M^{(k)} + m^{(k)}(t)}{M + m(t)} \tag{3.14}$$

The enrichment e is defined as the abundance of isotope $^{(k)}$ above its natural level

$$e(t) = a^{(k)}(t) - a_n^{(k)}(t) = \frac{M^{(k)} + m^{(k)}(t)}{M + m(t)} - \frac{M^{(k)}}{M} \tag{3.15}$$

From the expressions above, it is evident that none of the above variables is ttr, and that their use as measurement variable requires a formalism different from the one in use for radioactive tracer. Assume, for example, isotope ratio $r^{(k)}$ as defined by Equation 3.13 is the measurement variable. By simple algebra, the measurement equation can be written as a function of the state variable m and of tracee mass M as follows

$$r^{(k)}(t) = \frac{m(t)r_i^{(k)} + Mr_n^{(k)}\dfrac{1 + r_i^{(1)} + \ldots r_i^{(k)} + \ldots}{1 + r_n^{(1)} + \ldots r_n^{(k)} + \ldots}}{m(t) + M\dfrac{1 + r_i^{(1)} + \ldots r_i^{(k)} + \ldots}{1 + r_n^{(1)} + \ldots r_n^{(k)} + \ldots}} \tag{3.16}$$

where $r_n^{(1)} \ldots r_n^{(k)} \ldots$ and $r_i^{(1)} \ldots r_i^{(k)} \ldots$ are respectively the natural (of the tracee) and the tracer isotope ratios, e.g.,

$$r_n^{(1)} = \frac{M^{(1)}}{M^{(a)}} \ldots \qquad r_i^{(1)} = \frac{m^{(1)}}{m^{(a)}} \ldots$$

Equation 3.16 is a nonlinear function of m, where m appears both in the numerator and denominator of the measurement equation. Nonlinear functions similar to this also link abundance and enrichment.

Conversely, as discussed in Section 3.2, the use of tracer-to-tracee ratio

$$ttr(t) = \frac{m(t)}{M}$$

as measurement variable allows one to maintain the same formalism as for the radioactive case. However, since ttr is not directly measured by mass spectrometry, conversion methods must be applied. The expression, previously derived to calculate ttr from isotope ratios is the following

$$ttr = \frac{\mathrm{r}^{(k)} - \mathrm{r}_\mathrm{n}^{(k)}}{\mathrm{r}_\mathrm{i}^{(k)} - \mathrm{r}^{(k)}} \cdot \frac{1 + \mathrm{r}_1^{(1)} \ldots \mathrm{r}_\mathrm{i}^{(k)} + \ldots}{1 + \mathrm{r}_n^{(1)} \ldots \mathrm{r}_n^{(k)} + \ldots}. \tag{3.17}$$

Formulas are also available to derive ttr from other mass spectrometry variables such as abundance and enrichment.[11,12]

**Example – Single stable isotope tracer**

In Table 3.1, the natural abundance of zinc is reported, together with the abundances of two different zinc tracers used in Lowe et al., $^{67}$Zn and $^{70}$Zn.[8] Isotope ratios of tracee and of the two tracers, calculated by considering the isotope 66 as reference, are also shown.

**TABLE 3.1**

Zinc Isotope Abundance and Isotope Ratio of Tracee and Tracers

| | Abundance | | | Isotope Ratio | | |
|---|---|---|---|---|---|---|
| **Isotopes** | **Tracee** | **$^{67}$Zn Tracer** | **$^{70}$Zn Tracer** | **Tracee** | **$^{67}$Zn Tracer** | **$^{70}$Zn Tracer** |
| 64 | 0.4889 | 0.0111 | 0.0583 | 1.7580 | 0.4237 | 1.5423 |
| 66 | 0.2781 | 0.0262 | 0.0378 | 1.0000 | 1.0000 | 1.0000 |
| 67 | 0.0411 | 0.9180 | 0.0071 | 0.1478 | 35.0382 | 0.1878 |
| 68 | 0.1857 | 0.0444 | 0.0465 | 0.6677 | 1.6947 | 1.2302 |
| 70 | 0.0062 | 0.0003 | 0.8503 | 0.0223 | 0.0115 | 22.4947 |
| Sum | 1.0000 | 1.0000 | 1.0000 | 3.5958 | 38.1681 | 26.4550 |

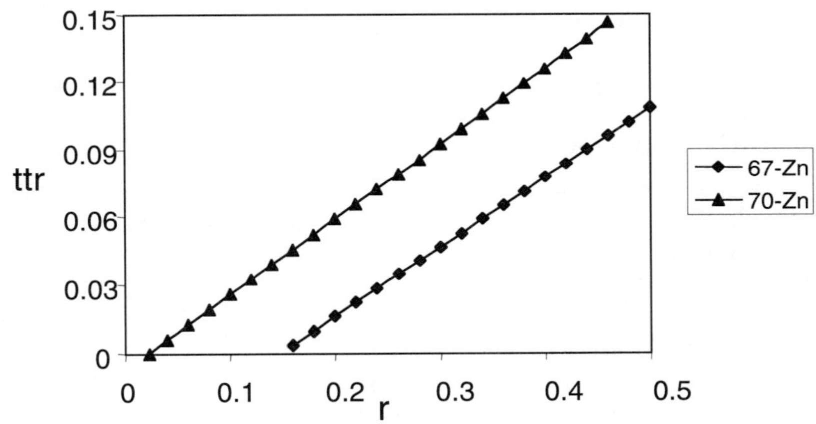


**FIGURE 3.2**
Relationship between isotope ratio r and tracer-to-tracee ratio ttr for two different zinc tracers.

Equation 3.17 allows one to express the link between isotope ratio and tracer-to-tracee ratio for each tracer:

$$^{67}\text{Zn tracer: } ttr = \frac{r^{(1)} - 0.1478}{35.0382 - r^{(1)}} 10.61463 = \frac{r^{(1)} - 0.1478}{3.3009 - 0.0942 r^{(1)}} \tag{3.18}$$

$$^{70}\text{Zn tracer: } ttr = \frac{r^{(4)} - 0.0223}{22.4947 - r^{(4)}} 7.3572 = \frac{r^{(4)} - 0.0223}{3.0572 - 0.1359 r^{(4)}} \tag{3.19}$$

The relationships between r and ttr for the two tracers for different values of r are shown in Figure 3.2. For example, if $^{67}$Zn is used as a tracer and the isotope ratio in a sample is 0.2145, then, from Equation 3.17, ttr in that sample is ttr = 0.020331.

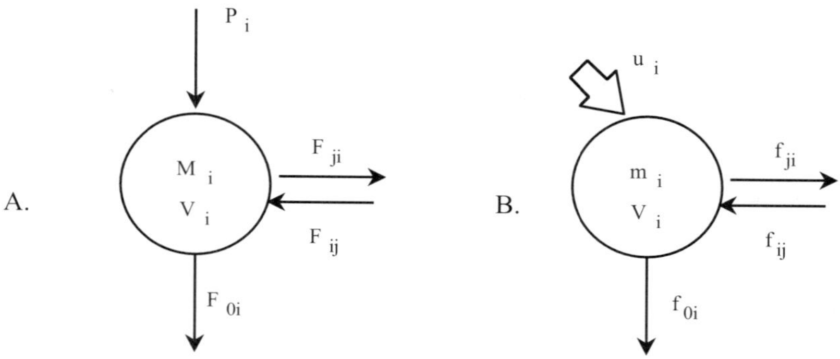


**FIGURE 3.3**
(A) The n-compartment tracee system. (B) The n-compartment tracer system.

## 3.4 Multi-pool Tracer Kinetics and Measurement

Although in Section 3.2 we have considered the single pool system, we assume now a multipool system in steady state and we will discuss state and measurement variables for compartmental modelling of tracer kinetics. A compartmental model consists of a finite number of compartments (representing collections of particles, at specific sites and/or in specific forms, which are homogeneous and behave identically) with specified interconnections among them (representing fluxes of material, e.g., transport from one location to another or chemical transformation or both). If the model consists of n compartments, and the tracee system is in steady state during the experiment, by using the mass balance principle one can write for each compartment (Figure 3.3)

$$\frac{dM_i}{dt} = -\sum_{\substack{j=0 \\ j\neq i}}^{n} F_{ji} + \sum_{\substack{j=1 \\ j\neq i}}^{n} F_{ji} + U_i = 0 \tag{3.20}$$

where $M_i$ is the tracee mass in compartment i, $F_{ij}$ is the flux (mass/time) from compartment j to compartment i, $F_{01}$ is disposal (mass/time) from compartment i, and $U_i$ (mass/time) *de novo* production into compartment i.

If the experiment consists of an ideal tracer input into the accessible compartment, e.g., compartment 1, mass balance equations, written in terms of the tracer masses in compartments, $m_i$, i = 1,2…n, are

$$\frac{dm_i(t)}{dt} = -\sum_{\substack{j=0 \\ j\neq i}}^{n} f_{ji}(t) + \sum_{\substack{j=1 \\ j\neq i}}^{n} f_{ij}(t) + u_i(t) \tag{3.21}$$

where $m_i$ is the tracer mass in compartment i, $f_{ij}$ is the tracer flux (mass/time) from compartment j to compartment i, $f_{01}$ is tracer disposal (mass/time) from compartment i, and $u_i$ (mass/time) is tracer input in compartment i, different from zero if i = 1. The link between the tracee and tracer system comes from the indistinguishability principle of tracer and tracee (Equation 3.3), now generalized as:

$$f_{ij}(t) = \frac{F_{ij}}{M_j} m_j(t) = k_{ij} m_j(t) \tag{3.22}$$

where $k_{ij}$ is constant since $f_{ij}$ and $M_j$ are constant. They have unit time$^{-1}$ and are called rate constants or fractional transfer coefficients. By using Equation 3.22 into Equation 3.21, tracer equations become

$$\frac{dm_i(t)}{dt} = -\sum_{\substack{j=0 \\ j \neq i}}^{n} k_{ji} m_i(t) + \sum_{\substack{j=1 \\ j \neq i}}^{n} k_{ij} m_j(t) + u_i(t) \tag{3.23}$$

In order to estimate the rate constants from tracer kinetics, one must link the equations describing the model with the tracer measurement variables. For a radioactive tracer, the measurement equation usually expresses the tracer concentration in the accessible compartment 1 as $c_1(t) = m_1(t)/V_1$, where $V_1$ is the distribution volume. Then, the state differential equations are linear in the variables $m_i$. Similarly, the measurement equation is a linear function of $m_1$.

For the stable-isotope tracer, it is convenient to formulate the system-experiment model equations in a similar way. Assuming the stable-isotope tracer masses in the compartments $m_i$, i = 1,2…n as state variables, one can write n linear mass balance equations just as in the radioactive tracer case. The measurement equation is linear if it is expressed in terms of the tracer-to-tracee ratio in the accessible compartment, $ttr_1(t) = m_1(t)/M_1$. This equation is very similar to the radioactive tracer measurement with tracee mass $M_1$ replacing the volume $V_1$. Conversely, as in the single pool case, there are other ways to express stable-isotope measurements, e.g., isotope ratio or enrichment result in a nonlinear function of the tracer mass in the accessible pool.

In summary, by using tracer-to-tracee ratio as the measurement variable, it is possible to formulate the compartmental model equations for the radioactive tracer and the stable-isotope tracer in a similar way. This is convenient because it allows one to extend to the stable isotope tracers case a number of results already proven for radioactive tracer compartmental models, such as *a priori* identifiability results and multiexponential modelling.

As far as *a priori* identifiability is concerned, a number of methods exists to test *a priori* identifiability of linear time invariant compartmental models.[2] They can be applied if the data are expressed as the tracer-to-tracee ratio, since in this case the output equation is linear in the state variables. From the similarity between the radioactive and stable-isotope tracer formalism, it is possible to test *a priori* identifiability of the system-experiment model irrespective of the tracer used, since state equations are the same for both tracers, and output equations are the same, except that the tracee mass appears in the output equation in place of the volume.

Some aspects of compartmental analysis can be profitably performed by fitting a sum of exponential model to the data, since it represents the solution of a system of linear differential equations. For example, using this procedure, one can determine the model order, i.e., the number of compartments in the system, which is equal to or at least not lower than the number of exponential terms which can be fitted accurately to a set of data. Multiexponential analysis is well established in radioactive studies; for stable-isotope tracers, the use of the tracer-to-tracee ratio, but not of isotope ratio or enrichment as the output variable, allows one to perform the multiexponential analysis in a similar way.

## 3.5 The Multiple Tracer Case

Up to this point, we have considered the case in which only one compartment is accessible for test input and measurements and a single tracer experiment is performed. More informative experiments can be performed when more than one compartment is accessible and/or different tracer inputs are used. For instance, to characterize the absorption of an element, a tracer can be injected into plasma and a second one given orally. If measurements are taken from plasma, such a study will result in two tracer responses. If compartments other than plasma (such as urine and feces) are accessible to measurement, the database for compartmental modelling consists of six tracer responses related to each tracer in the three accessible pools. However, it is important that the various tracer inputs are administered simultaneously to assure that the system is in the same condition for each and that the measurement procedures are able to quantitate the tracers separately.

The ideas discussed in Section 3.2 for the single tracer case in terms of the kinetic variables can be extended to the multiple tracer case. In particular, the measurement variables for radioactive tracers are the individual tracer concentrations. For stable isotopes, they are the tracer-to-tracee ratios, defined for the case where the two tracers are denoted as I and II and the measurement is taken from compartment 1 as:

$$ttr_1^I = \frac{\text{mass in compartment 1 from tracer I}}{\text{tracee mass in compartment 1}} \tag{3.24}$$

$$ttr_1^{II} = \frac{\text{mass in compartment 1 from tracer II}}{\text{tracee mass in compartment 1}}$$

If the model consists of n compartments, one can write n steady state equations for the tracee equal to Equation 3.20. For the tracers, two sets of n differential equations similar to Equation 3.23 can be written, one for each tracer, which differ because tracer inputs enter different compartments. For instance, if in an absorption study plasma is referred to as compartment 1 and the gastrointestinal tract as compartment 2, the differential equations describing the intravenous and the oral tracers have, respectively the tracer input in compartment 1 ($u_1$ different from zero) and 2 ($u_2$ different from zero). More output equations can be written, one for each measurement, and for each of them all the properties already mentioned apply. In particular, they are linear if the measurements are expressed as tracer-to-tracee ratios in accessible compartments.

Consider now the calculation of tracer-to-tracee ratios $ttr^I$ and $ttr^{II}$ in accessible compartments in terms of isotope ratio measurements. Expressions, first presented by Buckley[15], were derived for a dual-stable isotope study in Lowe et al. and are reported below for a generic compartment, since they are the same irrespective of the compartment from which the sample is taken[8]

$$ttr^I = \frac{[r^{(h)} - r_n^{(h)}] \cdot [r_j^{(k)} - r^{(k)}] - [r^{(k)} - r_n^{(k)}] \cdot [r_j^{(h)} - r^{(h)}]}{[r_i^{(h)} - r^{(h)}] \cdot [r_j^{(k)} - r^{(k)}] - [r_i^{(k)} - r^{(k)}] \cdot [r_j^{(h)} - r^{(h)}]} \cdot \frac{[1 + r_i^{(1)} \ldots]}{[1 + r_n^{(1)} \ldots]} \tag{3.25}$$

$$ttr^{II} = -\frac{[r^{(h)} - r_n^{(h)}] \cdot [r_i^{(k)} - r^{(k)}] - [r^{(k)} - r_n^{(k)}] \cdot [r_i^{(h)} - r^{(h)}]}{[r_i^{(k)} - r^{(h)}] \cdot [r_j^{(k)} - r^{(k)}] - [r_i^{(k)} - r^{(k)}] \cdot [r_j^{(h)} - r^{(h)}]} \cdot \frac{[1 + r_j^{(1)} \ldots]}{[1 + r_n^{(1)} \ldots]}$$

We indicated with suffix $^h$ and $^k$ ($h < k$) the isotopes artificially enriched in tracer I and II respectively, with $r_n^{(1)}$, $r_n^{(2)}$ ... natural isotope ratios, $r_i^{(1)}$, $r_i^{(2)}$ ..., and $r_j^{(1)}$, $r_j^{(2)}$ ... the isotope ratios of tracer I and II.

The extension of ttr formulas to the multiple (more than two) tracer which, although possible, leads to very complicated formulas, is made possible in practice by the TTRM software tool presented in Section 3.7.

**Example – Dual stable isotope tracers**

In Lowe et al., two zinc tracers highly enriched in $^{67}Zn$ and $^{70}Zn$, respectively, were simultaneously administered orally and intravenously to provide a database for developing a model of zinc metabolism in humans.[8] Values of tracee and tracer isotope ratios are the ones already reported in Table 3.1. By applying Equation 3.25, the expressions for $ttr^I$ and $ttr^{II}$ in terms of isotope ratio measurements were derived

$$ttr^I = \frac{r^{(4)} + 0.0003r^{(1)} - 0.0224}{3.0705 - 0.1358r^{(4)} - 0.0876r^{(1)}}$$

$$ttr^{II} = \frac{r^{(1)} + 0.0018r^{(4)} - 0.1478}{3.3042 - 0.1461r^{(4)} - 0.0943r^{(1)}} \tag{3.26}$$

## 3.6 A Test of the Endogenous-constant, Steady-state Assumption

The assumption that the endogenous constant steady state is not perturbed by the administration of the tracer is essential to write linear time invariant compartmental model equations. This assumption is not critical with radioactive tracers, since their mass is usually negligible compared to the tracee mass. With stable-isotope tracers, however, the tracer mass is often non-negligible: the tracee steady-state assumption is satisfied if tracee production is constant and equal to the pre-test value, and the system kinetics are linear in the range of values during the experiment. In most cases where the tracer perturbation is confined within a few percent, this range is usually small and the conditions above are likely to be satisfied. However, it is possible to test the steady-state assumption by a method which relies on measurements only, i.e., no assumptions about the system structure are required. The method is based on the additional measurement of total concentration, equal to tracer-plus-tracee concentration, which is usually available. Consider the single tracer case: by resorting to ttr definition one has

$$c_{\text{tot}}(t) = \frac{M(t) + m(t)}{V} = \frac{M(t)}{V}\left[1 + \frac{m(t)}{M(t)}\right] = C(t)[1 + ttr(1)] \qquad (3.27)$$

where the tracee concentration C is now considered as a function of time, to account for the possibility of a perturbation. From Equation 3.27 tracee concentration can be evaluated

$$C(t) = \frac{c_{\text{tot}}(t)}{1 + ttr(t)} \qquad (3.28)$$

Equation 3.28, which can be easily extended to the multiple tracer case, allows one to evaluate the extent to which the tracee concentration is perturbed from its pre-test constant steady-state value. The test may be a confirmatory one. Alternatively, it may suggest how to improve the experimental design (e.g., a reduction of the tracer dose, a more gentle input format) or the model (e.g., a structural description of feedback mechanisms or of nonlinear kinetics).

## 3.7 Software Tool: TTRM

The procedure to calculate the tracer-to-tracee ratio from primary mass spectrometry measurements for both the single-tracer case (Equation 3.17) and the dual-tracer case (Equation 3.25) as well as its generalization to multiple tracers has been implemented in a software tool, TTRM (available upon request). The program allows one to deal with the minerals most used in kinetic studies. Through a user-friendly interface, TTRM requires one to specify as input parameters:

1. The mineral.
2. The isotope used as the reference isotope.
3. Tracee masses of all isotopes, in arbitrary units, from a sample taken prior to the experiment; from these masses, isotope ratios $r_n^{(1)}$, $r_n^{(2)}$ ... are evaluated.
4. Tracer masses of all isotopes, in arbitrary units, from a sample of the infused material; from these masses, isotope ratios $r_i^{(1)}$, $r_i^{(2)}$ ... are evaluated. In the multiple-tracer case, point 4 is iterated for each tracer.
5. Masses in the experimental sample of the reference isotope and of the artificially enriched isotope(s), in arbitrary units, from which isotope ratio(s) $r^{(k)}$ (single tracer) or $r^{(k)}$, $r^{(h)}$ (two tracers), or $r^{(k)}$, $r^{(h)}$ ... (more than two tracers) are evaluated.

TTRM provides as output parameters the ttr value(s) in the sample.

An additional feature of TTRM is its evaluation of the precision and the accuracy of ttr; that is, the effect on ttr calculations of measurement errors. Random errors invariably present in the measurements affect ttr precision; it is evaluated via Monte Carlo methods.[14] The assumption is made that measurement errors are Gaussian, independent, zero mean, and with a known standard deviation which must be supplied by the user. The accuracy of ttr results from systematic errors which may affect measurements in non-ideal conditions, e.g., due to inter-day variability in the instrument response or isotope effects. These can only be corrected empirically, by means of an instrument calibration line which is built by TTRM based on additional measurements of standard samples.

**Example – Single stable isotope tracer**

Consider the data used in Table 3.1. If TTRM is used with the $^{67}Zn$ tracer, input parameters are

1. Zinc
2. 66
3. 1.7580 1. 0.1478 0.6677 0.0223
4. 0.4237 1. 35.0382 1.6947 0.0115
5. 0.2145

If the assumption is made that the error has a constant coefficient of variation (CV = 1%), one obtains the value of ttr in the sample, together with its precision: ttr = 0.020331, CV = 4%.

## 3.8 Conclusion

By using ttr as the measurement variable in compartmental modelling of stable-isotope tracer data, it is possible to establish a link with the kinetic formalism available for radioactive tracer data. Compartmental equations describing the system-experiment model when the endogenous system is in a constant steady state can be written in an analogous format for both radioactive and stable-isotope tracers. Since ttr is not directly measured by mass spectrometry techniques, a method has been presented to determine ttr in practice, both for the single- and the multiple-tracer case. Finally, with ttr, the assumption that the endogenous constant steady-state is not perturbed by the administration of the stable-isotope tracer can be tested from measurements only.

## References

1. Cobelli, C. and Caumo, A., Using what is accessible to measure that which is not: necessity of model of system, *Metabolism*, 47, 1009–1035, 1998.
2. Cobelli, C., Foster, D., and Toffolo, G., *Tracer Kinetic in Biomedical Research: From Data to Model*, Kluwer Academic/Plenum Publishers, New York, NY, in press.
3. Foster, D.M., Aamodt, R.L., Henkin, R.I., and Berman, M., Zinc metabolism in humans: a kinetic model, *Am. J. Physiol.*, 237, R340–R349, 1979.
4. Scott, K.C. and Turnlund, J.R., A compartmental model of zinc metabolism in adult men used to study effects of three levels of dietary copper, *Am. J. Physiol.* 267, E165–E173, 1994.
5. Wastney, M.E., Aamondt, R.L., Rumble, W.F., and Henkin, R.I., Kinetic analysis of zinc metabolism and its regulation in normal humans, *Am. J. Physiol.*, 251, R398, 1986.
6. Janghorbani, M. and Young, V.R., Advances in the use of stable isotopes of mineral in human studies, *Federation Proc.*, 41, 2702, 1982.
7. Wastney, M.E., Gökmen, I.G., Aamodt, R.L., Rumble, W.F., Goldon, G.E., and Henkin, R.I., Kinetic analysis of zinc metabolism in humans after simultaneous administration of $^{65}$Zn and $^{70}$Zn. *Am. J. Physiol.* 260, R134–R141, 1991.
8. Lowe, N.M., Shames, D.M., Woodhouse, L.R., Matel, J.S., Roehl, R., Saccomani, M.P., Toffolo, G., Cobelli, C., and King, J.C., A compartmental model of zinc metabolism in healthy women using oral and intravenous stable isotope tracers, *Am. J. Clin. Nutr.* 65, 1810–1819, 1997.
9. Patterson, B.H., and Zech, L.A., Development of a model for selenite metabolism in humans. *J. Nutr.*, 122, 709–714, 1992.
10. Turnlund, J.R., Human whole-body copper metabolism, *Am. J. Clin. Nutr., 67,* 960S–964S, 1998.

11. Cobelli, C., Toffolo, G., and Foster, D.M., Tracer-to-tracee ratio for analysis of stable isotope tracer data: link with radioactive kinetic formalism, *Am. J. Physiol.*, 262, E968, 1992.
12. Cobelli, C., Toffolo, G., Bier, D., and Nosadini, R., Models to interpret kinetic data in stable isotope tracer studies, *Am. J. Physiol.*, 253, E551, 1987.
13. Buckley, W.T., Huckin, S.N., and Eigendorf, G.K., Calculation of stable isotope enrichment for tracer kinetic procedures, *Biomed. Mass Spectrom.*, 12, 1–5, 1985.
14. Efron, B. and Tibshirani, R., Bootstrap methods for standard errors, confidence intervals, and other measures of statistical accuracy, *Stat. Sci.*, 1, 54–77, 1986.
15. Buckley, W.T., Budac, J.J., Godfrey, D.V., Koenig, K.M., Determination of multiple selenium stable isotope tracers by inductively coupled plasma spectrometry, *Biomed. Mass Spectrom.*, 21, 473–478, 1992.

# 4

# *Methods for Analysis of Trace-element Absorption*

**S.J. Fairweather-Tait, T.E. Fox, L.J. Harvey, B. Teucher, and J. Dainty**

**CONTENTS**

4.1 General Introduction ......60
4.1.1 Use of Isotopes ......60
4.1.2 Methods ......60
4.1.3 Definition of Absorption ......61
4.2 Iron ......62
4.2.1 Introduction ......62
4.2.2 Normalization of Iron Absorption Data ......62
4.2.3 Hemoglobin Incorporation ......63
4.2.4 Whole-body Counting ......64
4.2.5 Fecal Monitoring ......64
4.2.6 Plasma Appearance/Disappearance ......65
4.2.7 *In vitro* (Caco-2 Cells) ......65
4.2.8 Conclusion ......66
4.3 Copper ......66
4.3.1 Introduction ......66
4.3.2 Fecal Monitoring ......66
4.3.3 Plasma Appearance ......67
4.3.4 Whole-body Counting ......67
4.3.5 Conclusion ......68
4.4 Zinc ......68
4.4.1 Introduction ......68
4.4.2 Whole-body Counting ......68
4.4.3 Fecal Monitoring ......69
4.4.4 Urinary Monitoring ......70
4.4.5 Plasma Appearance/Disappearance ......70
4.4.6 Use of Simulation to Predict Absorption ......71
4.4.7 Whole-gut Lavage Technique ......71
4.4.8 *In vitro* (Caco-2 Cells) ......72
4.4.9 Conclusion ......72

0-8493-8730-2/00/$0.00+$.50

4.5 Selenium ..... 72
4.5.1 Introduction ..... 72
4.5.2 Fecal Monitoring ..... 74
4.5.3 Plasma Appearance/Disappearance ..... 74
4.5.4 Whole-body Counting ..... 75
4.5.5 Urinary Monitoring ..... 75
4.5.6 Conclusion ..... 76
References ..... 76

## 4.1 General Introduction

### 4.1.1 Use of Isotopes

Measuring trace-element absorption from the diet requires the use of isotopes to label the trace-element source. This allows differentiation between the proportion of trace element derived from the test food(s) and that derived from other sources (dietary or endogenous origin) in body fluids and tissues, as illustrated in Figure 4.1. The isotope used to label the food(s) must be present in the same chemical form as the native trace element. This can be achieved through biosynthetic (intrinsic) labelling, but it is an expensive and time-consuming approach; therefore, extrinsic labelling is used wherever possible. The latter technique was developed for iron using radioisotopes and is valid under conditions where complete isotopic exchange takes place prior to absorption.[1]

Radio- and stable isotopes can be used for multiple labelling studies for some trace elements (see sections below) and both offer a number of advantages and disadvantages. In general, radioisotopes are less costly and easier to measure than stable isotopes. More important, only a tracer quantity of the element is required for labelling purposes, whereas special problems arise with stable isotopes because they are naturally present in the environment and doses must be large enough to produce a measurable change in isotope ratios.[2] On the other hand, radioisotopes may have an inappropriate half-life and limited applicability due to ethical constraints or potential contamination problems. Samples from stable-isotope studies can be stored indefinitely but the analysis is very exacting.[3]

### 4.1.2 Methods

There are several methods that can be used to determine trace-element absorption; these are summarized in Table 4.1 and the relevant techniques discussed in detail under the sections on iron, zinc, copper, and selenium. Choice of method depends on the element under investigation, aim(s) of the study, location, characteristics of volunteers, and resources and skills available.

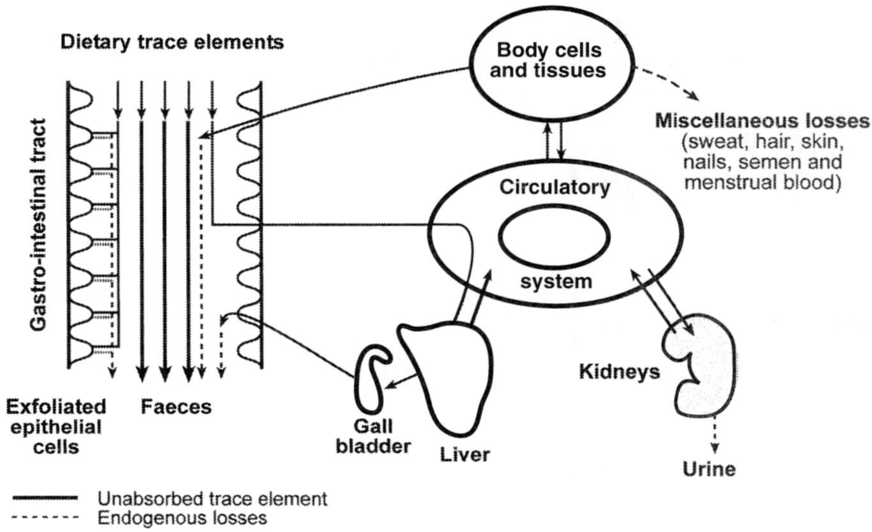


**FIGURE 4.1**
Trace element metabolism.

**TABLE 4.1**

| Approaches Used to Study Trace-Element Absorption Using Isotopes |
|---|
| Isotope balance (fecal plus urinary monitoring) |
| Whole body retention ($\gamma$-emitting isotopes) |
| Plasma appearance/disappearance (AUC, deconvolution) |
| Urinary excretion (double isotope technique) |
| Hemoglobin incorporation (iron) |
| *In vitro* techniques (Caco-2 cell systems) |

Broadly speaking, absorption can either be determined from luminal disappearance data or from monitoring the appearance of the isotope in body fluids or tissues.

### 4.1.3 Definition of Absorption

Absorption is a three-stage process comprised of uptake of the trace element from the GI lumen by the mucosal cells, intra-enterocyte transfer, and serosal transport into the systemic circulation. Any labelled element that is not transported into the body but is lost through mucosal cell exfoliation is usually not classified as absorbed. The difference between isotope intake and excretion (in feces and urine) is called *apparent absorption* and when an allowance is made for the quantity of isotope that is lost through endogenous excretions, the term *true absorption* is used. Whole-body counting techniques can either generate apparent or true absorption data, depending on the ability to quantify endogenous excretion. *Fractional absorption* is a term used to describe the proportion of the dose that has been absorbed, and *retention* describes the quantity of dose that is retained in the body after urinary and other losses have been taken into account.

When selecting an *in vivo* method for measuring trace-element absorption, a power calculation should be performed to estimate the number of volunteers needed for study such that the output measures will generate statistically significant results.[4]

## 4.2 Iron

### 4.2.1 Introduction

Iron is present in the diet as heme and non-heme iron, and because these are absorbed by independent pathways, they must be labelled separately. Iron homeostasis is maintained by varying the efficiency of absorption since there are no major excretory pathways for iron; thus, absorption is a key determinant of iron balance. Several radioisotope techniques are available to measure absorption, some of which have been adapted in recent years for use with stable isotopes.

One of the major problems with iron is the very wide inter- and intra-individual variation in iron absorption.[5] Adopting a multiple-dosing protocol can blunt day-to-day fluctuations in efficiency of absorption. Inter-individual variation is primarily due to differences in iron stores, as reflected by serum ferritin concentration. There is a significant inverse relationship between iron absorption and serum ferritin concentration, but not at very low or high (>60 μg/L) concentrations.[6] Changes in efficiency of absorption of non-hem iron are much more pronounced than with hem iron because the latter is better absorbed than non-hem iron. One of the challenges faced in designing iron absorption studies is to develop experimental protocols that enable results from different studies to be expressed in such a way that they can be compared and interpreted in the context of the relevant literature.

### 4.2.2 Normalization of Iron Absorption Data

Two main approaches are used to deal with the large inter-individual differences in iron absorption:

1. Normalization of data using well-absorbed reference dose (usually 3 mg iron as ferrous sulphate plus 30 mg ascorbic acid). Absorption from the test meal is corrected to a mean reference value of 40% by multiplying by 40/R, where R is the reference-dose absorption.[7]

2. Correction of absorption data to a value corresponding to a serum ferritin of 40 μg/L [8] using the equation

$$\text{Log } A_c = \text{Log } A_o + \text{Log } F_o - \text{Log } 40$$

where $A_c$ is the corrected absorption from the test meal, $A_o$ is the observed absorption, and $F_o$ is the observed serum ferritin.

These techniques are relevant to all *in vivo* methods but neither solves all the problems surrounding normalization of iron absorption data.

### 4.2.3 Hemoglobin Incorporation

Currently, the simplest and most widely used method for measuring iron absorption involves extrinsic labelling with a radioisotope of iron ($^{55}Fe$ or $^{59}Fe$), feeding one or more meals to fasting individuals, and measuring the percentage incorporation of the dose in the red blood cells 14 days after dosing.[9] A comparison can be made between two different sources of iron or a reference dose.[6,8] The very low levels of radioactivity (circa 37 kBq $^{59}Fe$ or 111 kBq $^{55}Fe$) necessitate careful sample preparation, using a modification of the method of Eakins and Brown.[10] Percentage iron absorption is calculated on the basis of blood volume (estimated from height and weight) and red cell incorporation of the absorbed dose (80% with serum ferritin >15 μg/L and 100% with serum ferritin <15 μg/L).[8] Alternatively, a dual-isotope method can be used in which $^{59}Fe$ is given orally and $^{55}Fe$ bound to plasma is simultaneously injected intravenously.[11] Absorption is calculated by relating the ratio of the two isotopes in red cells to the ratio of the administered isotopes. Whole-body counting can be used for verification (see section below).

Some countries are reluctant to permit the use of radioisotopes for nutrition research, particularly in women and children where unnecessary radiation exposure is not ethically acceptable. Thus the hemoglobin incorporation technique has been adapted for use with stable isotopes in infants, either as a single or double isotope technique.[12,13] Absorption is calculated by assuming that 90% of the absorbed iron is incorporated into hemoglobin. The dose of isotope required for the test depends on the anticipated absorption and the detection system whereby the limit of detection is three times the SD, and limit of quantification ten times the SD.[14,15]

Adapting the method for use in adults is more difficult because of the large doses of isotopes that have to be given for measureable enrichment of red blood cells (approximately ten times the quantity required for infant studies). This is costly and requires a multiple-dosing protocol. However, as improvements are made in mass spectrometric technologies, adult studies are becoming feasible, both technically and economically. The dual isotope technique has been used successfully to measure iron absorption from 5 mg $^{57}Fe$ (oral) and 250 μg $^{58}Fe$ (i.v.) by inductively coupled plasma mass spectrometry in women.[16] More recently, iron absorption by six-year-old Peruvian school children from two different school meals, labelled with 7 mg $^{57}$Fe or 2.6 mg $^{58}$Fe, was measured by the hemoglobin incorporation technique.[17] A further refinement to the method that will increase sensitivity is the measurement of isotope enrichment in reticulocytes, where newly absorbed isotope is found, instead of whole blood.[18]

### 4.2.4 Whole-body Counting

Single measurements of iron absorption can be performed using $^{59}$Fe and whole body counting.[19] Because the technique requires access to a whole body counter, it is not widely used. An initial baseline count is made to measure background activity from $^{40}$K and any residual radioactivity from previous studies involving radioisotopes. The volunteer is given a $^{59}$Fe-labelled source of iron and recounted one to five hours and 10 to 14 days later. The difference is taken to be absorbed isotope, once allowances are made for radioactive decay and geometry (i.e., point source immediately post-ingestion vs whole-body distribution 14 days later).

### 4.2.5 Fecal Monitoring

Radio- or stable-isotope balance can be used to measure iron absorption. Because there is no appreciable loss of absorbed isotope during the collection period (which may vary from 3 to 14 days), either via the urine or GI tract, apparent absorption equates to true absorption. Doses of stable isotopes are dependent on the detection system, but are generally less than those needed for the hemoglobin incorporation technique.[14] Unlike whole-body counting, multiple dosing can be employed to smooth out day-to-day variations in iron absorption, but there are constraints in relation to the length of the fecal collection period. Insufficient collection periods may lead to an overestimate of absorption if significant quantities of isotope have been taken up by mucosal cells but not transported into the body. This iron will be excreted when the cells are exfoliated, which will occur at a later time than excretion of unabsorbed isotope.

The most common problem with the fecal monitoring method is incomplete fecal collection, leading to an overestimate of absorption.[19] A number of fecal markers have been used to test for completeness of the collection with iron absorption studies; the most recent is rare earth elements.[20] Recovery of samarium, ytterbium, and dysprosium is close to 100% in adults, and these elements appear to follow the same excretory pattern as unabsorbed stable isotopes of iron. The situation is different in infants, however, where recoveries are lower (Fairweather-Tait, unpublished), suggesting that rare-earth elements are not suitable for assessing the completeness of fecal collections in infant studies.

### 4.2.6 Plasma Appearance/Disappearance

The absorption of $^{59}Fe$ measured by whole-body counting is highly correlated with serum iron increase 4 to 6 hours post-ingestion for large doses of iron (25–100 mg ferrous iron).[21] However, the technique cannot be used to measure iron absorption from levels of iron found in foods because of wide diurnal fluctuations in serum iron concentration.

Barrett et al. tested the validity of a post-absorptive serum stable-isotope enrichment method in which an oral dose of 2.8 mg $^{54}Fe$ (plus 10 mg $^{59}Fe$) was given simultaneously with an intravenous dose of 200 µg $^{57}Fe$ or $^{58}Fe$.[22] Stable isotope enrichment was measured in the plasma for ten hours as area under the curve (AUC) and oral absorption calculated as $(dose_{iv} \times AUC_{oral})/(AUC_{iv} \times dose_{oral})$. Absorption was also measured by whole-body counting. Although the mean results for the two methods were similar, there were wide differences between the two sets of data for several of the individuals. A further disadvantage is the fact that large volumes of blood need to be withdrawn for stable-isotope analysis, and this will affect body iron stores and hence limit the number of absorption tests that can be undertaken in any one individual.

The quantity of iron, its chemical form and rate of i.v. infusion must be carefully controlled.[23] In a recent dual-tracer experiment, which included a five to ten minute i.v. infusion of ferrous citrate, the calculations based on AUC predicted that the oral tracer was absorbed in excess of 100% of the initial dose (Fairweather-Tait, unpublished) – clearly an untenable situation.[22] On closer inspection of the data, it was observed that the i.v. iron was being cleared from the plasma at twice the rate of the oral iron, possibly due to the fact that the quantity of iron being infused was greater than the iron-binding capacity of plasma. This is a good example of the potential for misuse of the AUC calculation. The oral and i.v. forms of the mineral under investigation must behave identically once inside the body for the use of AUC to be valid; this can be checked by comparing the plasma-clearance rate of the two isotopes. Where there is a problem, it is possible to use plasma oral dose appearance data to estimate absorption, without correcting for rate of removal, by setting up a one-compartment iron kinetic model.[24]

### 4.2.7 *In vitro* (Caco-2 Cells)

*In vitro* methods do not measure iron absorption per se, but the Caco-2 cell system is worth mentioning as it is a useful predictive test for iron bioavailability. These cells are grown on microporous membranes in bicameral chambers where they differentiate spontaneously into bipolar enterocytes that exhibit many of the characteristics of normal epithelial cells. Radioisotopic tracers are used to label food sources to determine uptake and transport by the cell system and the results are consistent with *in vivo* data with respect to identifying differences in the bioavailability of iron compounds and predicting effects of enhancers and inhibitors.[25,26]

### 4.2.8 Conclusion

When radioisotopes are used, the method of choice would be hemoglobin incorporation or whole-body counting. Stable isotopes can be used in place of radioisotopes for the hemoglobin incorporation technique, provided the dose absorbed is high enough to enrich hemoglobin measurably. Plasma appearance/disappearance is only appropriate for situations when radioisotopes cannot be used or the doses of stable isotopes are not high enough to be detected. Fecal monitoring following stable isotope administration is an alternative method that requires lower doses of isotope and produces reliable results in experienced hands.

## 4.3 Copper

### 4.3.1 Introduction

Copper does not have a specific target tissue accessible for sampling in the human body where the uptake or retention of isotope could be used to assess bioavailability. Copper homeostasis in the human body is maintained by changes in both the absorptive efficiency in the gut and biliary copper excretion. Therefore, an estimate of endogenous copper losses is essential in order to measure true copper absorption.

The lack of appropriate radio- and stable isotopes of copper has precluded the use of some of the more elegant approaches such as dual labelling techniques used to determine the absorption of other minerals. The use of copper radioisotopes in metabolic research is restricted to only two of the seven available isotopes, $^{64}Cu$ and $^{67}Cu$, which have relatively short half-lives of 12.8 and 58.5 h, respectively. Consequently, the use of radioisotopes has been limited to short-term studies. The introduction of stable-isotope tracers has greatly facilitated research on copper absorption, allowing longer-term studies to be carried out in volunteers without exposure to radioactivity. However, the two stable isotopes of copper, $^{63}Cu$ and $^{65}Cu$ are relatively abundant, with natural abundances of 69.2% and 30.8%, respectively. The stable isotope of choice for copper absorption measurements is $^{65}Cu$.

### 4.3.2 Fecal Monitoring

Currently, the most widely used technique for assessing copper absorption is fecal monitoring. Both radio- and stable isotopes can be used, but the disadvantages of the former (short half-life, limited availability, and concern over safety of ionizing radiation) greatly restrict their use. Assessment of copper absorption in human volunteers has been made in several studies, with $^{65}Cu$, using doses corresponding from 50 to 200% of typical daily intakes in most studies.[27–29] In such studies, unabsorbed copper is generally found to be excreted in the feces within 5–7 days, followed by smaller amounts of endogenous excretion of the dose. Therefore, the period of time over which the study is performed is critical. The collection of individual fecal samples is also essential in order to obtain information on endogenous losses. In the case of copper, in addition to using rare-earth elements to test the completeness of fecal collection these elements can also be used to delineate unabsorbed isotope from endogenous secretions.[20,30] It should be noted that only negligible amounts of copper are lost in the urine and, consequently, this route of excretion does not usually need to be considered in copper absorption studies in healthy individuals.

There are several reports in which oral doses of $^{64}Cu$ are used to assess copper absorption. However, permissible radioisotope doses only allow the appearance in feces to be measured for up to 5 days. Therefore, in order to calculate true copper absorption, an i.v. dose of $^{64}Cu$ should be given on a different occasion to estimate endogenous losses.[31,32]

### 4.3.3 Plasma Appearance

Currently, there is no quantitative method for assessing copper absorption using plasma appearance techniques. After copper has been transported into the body through the intestinal mucosal cells, it binds principally to albumin with a biological half-life of about 10 minutes. With such a fast turnover, albumin-bound copper may be a good candidate for monitoring changes in copper absorption. Techniques are currently under development for measuring $^{65}Cu$-bound albumin in plasma following an oral dose of stable isotope (Beattie and Harvey, unpublished).

### 4.3.4 Whole-body Counting

Whole-body counting is rarely used to measure copper absorption in healthy human volunteers, mainly because of the short half-life of the copper radioisotopes. $^{67}Cu$ has been employed to investigate copper absorption and retention in patients with genetic disorders of copper metabolism such as Wilson's disease.[33,34] The technique is similar to that used for assessing iron and zinc absorption.[35] The volunteer is given a $^{67}Cu$-labelled source of copper which is counted approximately 2 h after ingestion and then at regular intervals for a period of 2 to 3 weeks. Absorption is calculated as the y intercept of a semi-logarithmic plot of the percentage of $^{67}Cu$ retention (corrected for decay) against time after administration of the label.[36] The semi-logarithmic plot is usually linear from four to five days after the isotope is given. Failure to correct for interferences in the gamma-ray spectra from $^{214}Bi$ and $^{214}Pb$ arising from the decay of $^{222}Rn$ will lead to overestimation of $^{40}K$ baseline activity.

Additionally, it is essential that the initial $^{67}Cu$ dose is sufficient when measuring over an extended period.

### 4.3.5 Conclusion

Due to the problems associated with using copper radio isotopes, the method of choice for assessing copper absorption in human volunteers is fecal monitoring using stable-isotope methodologies. It is hoped that, in the future, a quantitative plasma appearance method will be developed to replace the fecal monitoring technique.

## 4.4 Zinc

### 4.4.1 Introduction

Zinc homeostasis is regulated by changes in the efficiency of absorption and the level of intestinal endogenous excretion.[37] The majority of investigators employ either a radio- or stable-isotope approach to measure absorption and endogenous excretion simultaneously (true absorption). There are two gamma-emitting zinc radioisotopes, $^{65}Zn$ and $^{69m}Zn$, with a physical half-life of 243.6 days and 13.9 hours, respectively. However, $^{65}Zn$ is usually the radioisotope of choice as $^{69m}Zn$ is costly and has limited commercial availability. Three out of the five zinc stable isotopes are of low enough abundance to be used as tracers in human research, i.e., $^{67}Zn$ (4.1%), $^{68}Zn$ (18.8%), and $^{70}Zn$ (0.6%). In recent years, zinc metabolic studies have predominantly been performed using zinc stable-isotope techniques because these can be applied to all population groups including children, women of child-bearing age, and pregnant and lactating women.

### 4.4.2 Whole-body Counting

Measurements of true zinc absorption can be made using $^{65}Zn$ and $^{m69}Zn$ and whole-body counting if corrections for background activity and re-excretion of absorbed zinc are made. The volunteer is given a $^{m69}Zn$-labelled test meal and whole-body counts (including $^{40}K$ baseline counts) are obtained within 1 hour of test meal consumption, and at regular intervals for periods of up to 14 days. Absorption is then corrected for re-excretion of zinc by measuring the excretion of an i.v. dose of $^{m69}Zn$ approximately 14 days after the oral dose.[38]

Protocols for determining zinc absorption using $^{65}Zn$ involve measuring whole-body retention 10 to 14 days after consumption of the test meal(s) to allow for excretion of the unabsorbed $^{65}Zn$. The long half-life of $^{65}Zn$ prohibits the use of a dual stable-isotope design where oral and i.v. doses are given to the same subject; thus a technique developed by Arvidsson et al. is widely used to measure true zinc absorption.[39] This corrects for re-excretion by using an average rate of excretion of an i.v. administered dose of $^{65}Zn$ from a different study group with similar characteristics. However, according to Wastney et al., re-excretion is characterized by a coefficient of variation of about 25%.[40] Thus correction for re-excretion may introduce a considerable error in calculating true absorption for an individual. A more time-consuming technique, based on regression analysis of the whole-body counts obtained between 7 and 60 days after administration of the oral test dose, is an alternative. Extrapolation of this curve to zero time gives the actual absorption of $^{65}Zn$.[41] This latter method is, however, not without disadvantage as retention may be affected by changes in dietary zinc intake over the long measurement period.

### 4.4.3 Fecal Monitoring

A $^{65}Zn$ radioisotope technique was developed to determine zinc absorption from the ratio of $^{65}Zn$ and a non-absorbed radioactive marker ($^{51}Cr$) present in the first single stool post-dosing that contains more than 10% $^{51}Cr$.[41,42] It is necessary to promote defecation by administering a laxative (such as lactulose syrup) approximately 8 hours after the test meal to ensure that unabsorbed radioactivity is promptly excreted. Large-volume gamma counters are employed to measure the activity of $^{65}Zn$ and $^{51}Cr$ in single stool samples. Since $^{51}Cr$ has been shown to follow the same excretory pattern as unabsorbed $^{65}Zn$, absorption can be calculated using the $^{51}Cr$ recovery to correct for incomplete recovery.[41] This method for measuring apparent absorption has been validated against whole-body counting.[41,42] The advantage of this technique lies in the short time required to complete the experiment and the lack of need for laborious sample processing. However, major disadvantages are the limitation to single meal experiments and the inability to measure endogenous losses directly.

Both single and dual stable-isotope techniques have been applied to studies using fecal monitoring to estimate true absorption. A minimum fecal collection period of 12 days is required when absorption is determined using the single stable-isotope approach. To correct for re-excretion of the absorbed tracer, cumulative isotope excretion (expressed as percent of administered dose) is plotted against time. Once all the unabsorbed isotope has been excreted, any further isotope appearing in the feces is of endogenous origin. A line is fitted by linear regression and extrapolated back to the y-axis, and true absorption calculated from the intercept.[43,44]

Using a dual stable-isotope technique, the luminal disappearance of the isotopes can be monitored following single test meals. True absorption is then determined by measuring the appearance of a second zinc stable isotope given intravenously to correct for intestinal re-secretion of absorbed isotope.[45] Knudsen et al. validated the stable-isotope fecal monitoring technique by comparing results with whole-body retention data and observed that fecal monitoring with stable zinc isotopes showed a larger variation.[46] This can largely be attributed to the multiple steps in sample processing prior to analysis by mass spectrometry. The fecal monitoring method requires complete collection of individual stools over several days (10 to 12 days). Compliance, one of the major concerns in relation to fecal monitoring, can be monitored by labelling the test meal with a non-absorbable rare-earth element, e.g., dysprosium chloride; this fecal marker has been shown to follow the same excretory pattern as unabsorbed zinc.[47]

### 4.4.4 Urinary Monitoring

Fractional zinc absorption can be determined following simultaneous oral and i.v. administration of different stable isotopes of zinc.[48] This method requires that 24-hour urine samples are collected from the time of isotope administration until the rate of disappearance, plotted against time, is the same for the oral and i.v. stable-isotope tracers. This usually occurs between 24 and 48 hours post-dosing. The method, however, is only applicable to single-meal studies. Furthermore, the amount of oral stable-isotope administered may become crucial when absorption and urinary excretion are low. Under such conditions, the isotope enrichment in urine may be below the levels of quantification of some mass spectrometry techniques. So far, the method has not been adequately validated. The finding that fractional absorption calculated from plasma appearance/disappearance agrees with that of urinary monitoring is not conclusive proof, as it has not yet been established that an i.v. infusion of zinc causes no perturbation to systemic zinc metabolism. (See Section 4.4.5.)

The total urinary zinc excretion following doses in excess of 25 mg of zinc is approximately linear. Thus urinary excretion may be considered as an alternative method of assessing absorption from zinc supplements.[49]

### 4.4.5 Plasma Appearance/Disappearance

The plasma zinc increase following a high dose of zinc (unlabelled) has been used to measure zinc bioavailability. However, it is generally agreed that the application of this method is restricted to either evaluating inter-individual zinc uptake from a test meal or supplements that contain higher zinc doses than those encountered in typical meals.[50–52]

The measurement of the appearance of $^{69m}Zn$ and zinc stable-isotopes in plasma following oral and i.v. administration can be used to measure fractional absorption.[38] While it is possible to administer stable isotopes simultaneously, i.v. and oral doses of $^{69m}Zn$ must be administered 2 weeks apart, although the use of $^{69m}Zn$ is rare for the reasons outlined above.[53] Assuming that i.v. and oral tracer undergo the same rate of removal from the plasma, the fractional absorption of the oral dose can be calculated from the dose-corrected oral/i.v. ratio of the areas under the plasma concentration vs time curve (AUC). It is noteworthy that the underlying assumption of same rate of clearance from the plasma has not been shown. This may be of special importance when using stable isotopes. Scott and Turnlund used compartmental modelling to demonstrate that zinc metabolism differed according to the route of administration of the stable-isotope tracers.[54] Both the relatively large amounts of stable isotope administered (0.2 to 2.0 mg) and the chemical form of the i.v. dose (usually citrate or sulphate) may change plasma kinetics and produce elevated absorption values. (See Section 4.2.) As with the urinary monitoring, validation of the method is essential and application of the technique is restricted to single-meal experiments.

### 4.4.6 Use of Simulation to Predict Absorption

One of the problems in absorption studies is knowing how much tracer (oral and/or i.v.) to give to a subject, bearing in mind the sensitivity of the mass spectrometer for measuring certain zinc ratios in biological samples. A simulation of an experimental protocol with, for example, SAAMII (SAAM Institute, Inc., Seattle, WA, U.S.), can be performed quickly and easily using an already published compartmental model.[55] Wastney et al. and Foster et al. derived models from radioisotope data.[40,56] Lowe et al. also based their model on the radioisotope models, but adapted it for stable isotopes by reducing the number of simulated body compartments.[57] By varying both the size of simulated tracer dose(s) and the simulated absorption, the masses of tracer zinc appearing in pools of interest over time can be predicted. This is extremely useful for calculating what the smallest appropriate tracer dose should be, given a very low efficiency of absorption, especially if the main interest is in endogenous losses or urinary excretion, where tracers are likely to be present in very small amounts.

Simulated mass spectrometer data can also be generated from the compartmental simulations. Noise can be added to these data from knowledge of the precision (RSD) of the instrument that will measure the real samples. This allows investigators to simulate when the potential limit of detection (LOD) and limit of quantification (LOQ) may occur in the sampling procedure. Knowledge of the time required to reach the LOQ and LOD should prevent the collection of unnecessary fecal or urine samples.

### 4.4.7 Whole-gut Lavage Technique

The whole-gut lavage technique has been used to study the bioavailability of various minerals.[58–62] The technique involves washing out the gastrointestinal tract with a large volume (~4L) of iso-osmotic solution by infusion through a tube. A test meal containing the non-absorbable marker PEG is then consumed. After an overnight fast, during which time absorption of zinc from the meal takes place, another intestinal washout is carried out to recover the unabsorbed portion of the test meal. Completeness of collection is determined from the recovery of the non-absorbable marker PEG. On a separate occasion, basal conditions are established by washing out the gut after a drink of water; zinc in the effluent represents endogenous excretion. True absorption is then calculated as the difference between the amount of unabsorbed zinc recovered in the effluent, corrected for the endogenous losses measured after taking water alone.[61] This technique has several drawbacks, including unpleasant side effects such as nausea and abdominal fullness. Also, it is unlikely that endogenous losses under basal conditions reflect the actual situation when food is ingested because water will not stimulate secretion of pancreatic fluid in the same way as food. Furthermore, endogenous zinc and dietary zinc may compete for absorption and this cannot be detected with this technique.[44]

### 4.4.8 *In vitro* (Caco-2 Cells)

In relation to zinc absorption, the Caco-2 cell system has recently been used to study: 1) the uptake mechanism of zinc at the apical and basolateral membrane; 2) zinc and copper interactions; and 3) the effect of caseinophosphopeptides on zinc absorption.[63–65] The principle of this *in vitro* technique is explained in the section on iron in this chapter. Studies measuring the uptake and transport of zinc from radioisotopically labelled food sources may provide predictive data for zinc bioavailability from various diets.

### 4.4.9 Conclusion

True zinc absorption from a single food/meal can be measured most accurately by performing whole-body counting for up to 60 days after administration of a radio-labelled test meal. However, disadvantages include the study length and maintaining consistency in dietary zinc intake during the study period. Apparent zinc absorption can be determined by counting for a shorter period, but this does not take into account endogenous losses. Urinary monitoring is the least invasive and most convenient stable-isotope technique, but the most important assumption concerning the validity of the method (i.e., same rate of clearance for the oral and i.v. dose) has not been unequivocally established. Furthermore, this technique is limited when the efficiency of absorption is low since isotope enrichment in the urine may be below the levels of quantification for some mass spectrometers.

## 4.5 Selenium

### 4.5.1 Introduction

The predominant forms of selenium in food are organically bound, primarily selenomethionine and other selenoamino acids and their methyl derivatives.[66]

A minor quantity of inorganic selenium is found in drinking water, the level depending on the selenium concentration of the soil and the soil type through which the water has passed. Larger intakes of inorganic selenium only occur when selenium supplements of selenate or selenite are taken.

There appears to be no homeostatic control of absorption of selenium and, under normal feeding conditions, absorption is not a limiting factor to bioavailability.[67–70] Selenite and selenate are absorbed passively across the brush border and compete with inorganic sulphur compounds.[71] Selenate absorption is assisted by a sodium pump, which explains the high absorption (>90%, similar to selenomethionine), whereas selenite has a lower apparent absorption, ranging from 30 to 60%.[72–76] Selenomethionine absorption is an active process using the same enzyme system as methionine in competition with its seleno analogue.[72,75] Similarly, selenocysteine competes with cysteine for an active transport system, while it is thought that other seleno-amino acids are absorbed passively.[78] Thus selenium does not form a common pool within the gut but several distinct pools with different absorption characteristics. The chemical form of a stable-isotope tracer used in a metabolic study will therefore be representative of only one of these intra-luminal pools. Studies employing isotopic tracers to investigate selenium absorption from foods must therefore use labels that have been biosynthetically incorporated during growth.

The systemic metabolism of absorbed selenium depends on the chemical form. Selenomethionine is largely taken up by the liver, where it is used for general protein synthesis because there is no metabolic distinction between the sulphur and amino analogues.[74,79] With selenite, only a minor fraction is removed from the hepatic portal circulation by the liver and a larger fraction is excreted in the urine.[80] Selenite is more readily available for selenoprotein synthesis, while selenomethionine is available only after catabolism of the protein into which it was incorporated. There are differences in the metabolism of selenite and selenate; the urinary clearance of absorbed selenate is faster than selenite even when an allowance is made for the lower efficiency of absorption exhibited by selenite.[68,73]

The efficiency of absorption of the seleno-amino acids and selenate is high and is remarkably consistent, with very little variability within or among individuals. This reflects the absence of homeostatic mechanisms controlling selenium absorption.[69,70,81] Selenite is more variable and affected by the food matrix with which it is consumed.

Selenium has six naturally occurring isotopes: $^{74}Se$, $^{76}Se$, $^{77}Se$, $^{78}Se$, $^{80}Se$ and $^{82}Se$. The natural abundance of all the isotopes, with the exception of $^{80}Se$, is low enough to allow them to be used as labels in metabolic studies employing enriched sources of the isotopes. Analysis of selenium in biological material is beset with problems leading to underestimation, with losses occurring due to difficulties in extracting selenium from organic matrices (e.g., trimethylselenonium excreted in the urine) plus the volatility of Se(IV) found in $SeO_2$ and organic compounds. Another problem encountered when analyzing Se by inductively coupled mass spectrometry is that the isotope

$^{80}Se$ has the same mass as the argon–argon diamer in the plasma source, which makes it impossible to measure the most abundant, naturally occurring selenium isotope. In this case, the full isotopic composition of all the isotopes in an enriched dose must be known in order to quantify the tracer in enriched samples.

### 4.5.2 Fecal Monitoring

The main excretory route for selenium is via the urine with very little loss through biliary and intestinal secretions in feces; thus apparent absorption is similar to true absorption figures.[82] Labelled oral doses of 80 to 200 μg selenium, which represent up to four times the daily intake, have been used in human studies.[81,83–85] Smaller doses of 60 μg, which are more representative of dietary intakes, could be used, providing care is exercised when processing fecal material and mass spectrometry measurements are not subject to large baseline fluctuations. Generally speaking, stable isotope doses of 100 μg with enrichments >70 atom % should be easily measurable in fecal samples. Collection of fecal samples must start immediately after the first dose and continue for at least 7 days after the last dose (if multiple dosing is used). Digestion of the fecal samples requires care in order to prevent loss of volatile compounds of selenium. Microwave digestion is commonly used, though open-vessel wet acid digestion can be used, providing the temperature does not exceed 180°C. To ensure complete fecal collection of unabsorbed label, a non-absorbable rare-earth element such as dysprosium chloride can be given with the selenium dose and the fecal samples monitored for complete recovery.

The only useful radionuclide of selenium for metabolic studies is $^{75}Se$, a gamma-emitting isotope without the presence of beta particles, with a half-life of 119.8 days. Radionuclides of selenium have been used to study selenium metabolism in humans using $^{75}Se$ as $^{75}Se$-selenomethionine[86] or $^{75}Se$-selenite.[87] In these studies, cumulative fecal losses of an oral dose supplying 20 μCi and 10 μCi, respectively, were measured in a large volume counter and absorption calculated. Endogenous losses of the absorbed dose can be estimated by plotting the cumulative fecal loss of the isotope against time over 10 to 14 days and extrapolating the linear section of the graph back to time zero. It should be noted that selenium accumulates in the testes of men and therefore would not be suitable for use with male subjects due to the increased health risks.

### 4.5.3 Plasma Appearance/Disappearance

As pointed out earlier, plasma kinetics cannot be used to measure selenium absorption from foods, due to the rapid removal of dietary selenium by the liver. Selenite undergoes changes by the enterocytes before it is released into the general circulation; there is a long lag phase before the appearance of selenite in the plasma, which precludes this method for assessing selenite absorption.

### 4.5.4 Whole-body Counting

Doses of 20 μCi $^{75}$Se-selenomethionine, delivering 130 to 168 mrad whole-body radiation and 200 mrad gonad dose, have been used successfully with whole-body counting to quantify the absorbed selenium. Selenite sources of $^{75}$Se will have smaller radiation doses because of the shorter biological half-life compared with selenomethionine sources.

### 4.5.5 Urinary Monitoring

True absorption can be assessed from urinary measurements of two tracers after a simultaneous labelled oral and i.v. dose for some minerals such as calcium.[88] The technique relies on complete exchange between the oral and i.v. tracer within different compartments of the body so that they are metabolized at the same rate and the urinary clearance of the two tracers is the same. This method has applications for selenite and selenate, providing that the oral and i.v. chemical forms are the same. Martin and colleagues found that, when using labelled selenite, the urinary and plasma enrichments equilibrated after 20 hours.[83] This finding suggests that, as with zinc, the technique can only be used with urine samples that have been collected two days after the administration of the dose. Absorption from the oral dose can then be measured in a single spot of urine according to the equation:

$$\%\text{Absorption} = \frac{\text{na}\,^{82}\text{Se}}{\text{na}\,^{74}\text{Se}} \times \frac{^{74}\text{Se(i.v.)}}{^{82}\text{Se(oral)}} \times \frac{\%\text{XS}\,^{82}\text{Se}}{\%\text{XS}\,^{74}\text{Se}}$$

where: na = natural abundance of the isotope (atom%)
i.v. and oral = doses of isotope administered (mM)
%XS = [(measured ratio – natural abundance ratio)/natural abundance] × 100

This application cannot be used for dietary forms of selenium found in foods since there are several distinct forms of selenium in the food and the method would require organic forms of selenium to be given intravenously, which is not possible. Great care must be taken, however, when using the above equation. Calculations involving tracers with low enrichment, possibly as a result of intrinsic labelling, will give incorrect results as a consequence of the inherent assumption in the %XS equation. Investigators should also be careful when only taking a spot urine sample since the measurement uncertainty may be quite high if the quantity of the oral and i.v. tracers in the sample is low. Bulked or cumulative samples will ensure lower measurement uncertainty because of their higher tracer content.

### 4.5.6 Conclusion

Estimates of selenium absorption from foods are limited to biosynthetically labelled foods with isotope tracers. Stable isotopes are more suited to this method, although $^{75}Se$, with its relatively short half-life, could be used where growth of the fruiting body of the plant occurs in under six weeks. At present, fecal monitoring of isotopic tracers or whole-body counting following $^{75}Se$ administration are the only options available for measuring the absorption of selenium in food. Estimates of inorganic selenium absorption are only representative of the absorption of selenium supplements of the same chemical form, since the absorption and metabolism of selenite and selenate differ. Fecal and urinary monitoring techniques can both be used to measure inorganic selenium absorption.

## References

1. Cook, J.D. et al., Food iron absorption measured by an extrinsic tag, *Journal of Clinical Investigation*, 51, 805, 1972.
2. Fairweather-Tait, S. J. and Fox, T. E., Intrinsic and extrinsic labelling of inorganic nutrients in food studies, *Stable Isotopes in Human Nutrition*, Mellon, F. A. and Sandström, B., Academic Press, London, 1996, 15.
3. Fairweather-Tait, S. J., The importance of trace element speciation in nutritional sciences, *Fresenius Journal of Analytical Chemistry*, 363, 536, 1999.
4. Bland, M., *An Introduction to Medical Statistics, Second Edition*, Oxford University Press, Oxford, 1995.
5. Kuhn, I.N. et al., Iron absorption in man, *Journal of Laboratory and Clinical Medicine*, 71, 715, 1968.
6. Hallberg, L., Hulten, L., and Gramatkovski, E., Iron absorption from the whole diet in men: how effective is the regulation of iron absorption?, *American Journal of Clinical Nutrition*, 66, 347, 1997.
7. Magnusson, B. et al., Iron absorption in relation to iron status. Model proposed to express results to food iron absorption measurements, *Scandinavian Journal of Haematology*, 27, 201, 1981.
8. Cook, J.D., Dassenko, S.A., and Lynch, S.R., Assessment of the role of nonheme-iron availability in iron balance, *American Journal of Clinical Nutrition*, 54, 717, 1991.
9. Brise, H. and Hallberg, L., A method for comparative studies on iron absorption in man using two radioiron isotopes, *Acta Medica Scandinavica Supplement*, 376, 59, 7, 1962.
10. Eakins, J.D. and Brown, D.A., An improved method for the simultaneous determination of iron-55 and iron-59 in blood by liquid scintillation counting, *International Journal of Applied Radiation and Isotopes*, 17, 391, 1966.

11. Saylor, L. and Finch, C.A., Determination of iron absorption using two isotopes of iron, *American Journal of Physiology*, 172, 372, 1953.
12. Fomon, S.T. et al., Erythrocyte incorporation of ingested $^{58}Fe$ by infants, *Pediatric Research*, 24, 20, 1988.
13. Kastenmayer, P. et al., A double stable isotope technique for measuring iron absorption in infants, *British Journal of Nutrition*, 71, 411, 1994.
14. van Dokkum, W. et al., Study techniques, *Stable Isotopes in Human Nutrition*, Mellon, F.A. and Sandström, B., Academic Press, London, 1996, 23.
15. Miller, J.C. and Miller, J.N., *Statistics for Analytical Chemistry*, 2nd edition, Ellis Horwood, New York, 1988.
16. Barrett, J.F.R. et al., Absorption of non-haem iron in normal women measured by the incorporation of two stable isotopes into erythrocytes, *Clinical Science*, 83, 213, 1992.
17. Walczyk, T. et al., Stable isotope labels as a tool to determine the iron absorption by Peruvian school children from a breakfast meal, *Fresenius Journal of Analytical Chemistry*, 359, 445, 1997.
18. van den Heuvel, E.G.H.M. et al., A new method to measure iron absorption from the enrichment of $^{57}Fe$ and $^{58}Fe$ in young erythroid cells, *Clinical Chemistry* 44, 649, 1998.
19. Bothwell, T.H. et al., *Iron Metabolism in Man*, Blackwell Scientific Publications, Oxford, 1979, 425.
20. Fairweather-Tait, S. J. et al., Rare earth elements as nonabsorbable fecal markers in studies of iron absorption, *American Journal of Clinical Nutrition*, 65, 970, 1997.
21. Ekenved, G., Norrby, A. and Solvell, L., Serum iron increase as a measure of iron absorption – studies on the correlation with total absorption, *Scandinavian Journal of Haematology Suppl.*, 28, 31, 1976.
22. Barrett, J.F.R. et al., Comparison of stable isotopes and radioisotopes in the measurement of iron absorption in healthy women, *Clinical Science*, 87, 91, 1994.
23. Veall, N. and Vetter, H., *Radioisotope Techniques in Clinical Research and Diagnosis*, Butterworth & Co., London, 1958, 254–271.
24. Gibaldi, M. and Perrier, D., *Pharmacokinetics*, 2nd edition, 1982, Dekker, New York, 1982, 145–198.
25. Garcia, M.N., Flowers, C., and Cook, J.D., The Caco-2 cell culture system can be used as a model to study food iron availability, *Journal of Nutrition*, 126, 251, 1996.
26. Glahn, R.P. et al., Caco-2 cell iron uptake from meat and casein digests parallels in vivo studies: use of a novel in vitro method for rapid estimation of iron bioavailability, *Journal of Nutrition*, 126, 332, 1996.
27. August, D., Janghorbani, M., and Young, V.R., Determination of zinc and copper absorption at three different Zn-Cu ratios by using stable isotope methods in young and elderly subjects, *American Journal of Clinical Nutrition*, 50, 1457, 1989.
28. Turnlund, J.R. et al., A stable isotope study of zinc, copper and iron absorption and retention by young women fed vitamin B-6 deficient diets, *American Journal Clinical Nutrition*, 54, 1059,1991.
29. Turnlund, J.R. et al., Copper absorption, excretion and retention by young men consuming low dietary copper determined by using the stable isotope $^{65}Cu$, *American Journal Clinical Nutrition*, 67, 1219, 1998.
30. Schuette, S.A. et al., Dysprosium as a nonabsorbable marker for studies of mineral absorption with stable isotope tracers in human subjects, *Journal of American College of Nutrition*, 12, 307, 1993.

31. Janssens, A.R., and van den Hamer, C.J.A., Kinetics of $^{64}$copper in primary biliary cirrhosis. *American Association for the Study of Liver Diseases*, 2, 822, 1982.
32. Bush, J.A. et al., Studies on copper metabolism. XVI. Radioactive copper studies in normal subjects and in patients with hepatolenticular degeneration, *Journal of Clinical Investigation*, 34, 1766, 1955.
33. Gunther, K. et al., The kinetics of copper uptake by the liver in Wilson's disease studied by a whole-body counter and a double labelling technique, *European Neurology*, 13, 385, 1975.
34. Sargent, T. and Stauffer, H., Whole-body counting of retention of $^{67}$Cu, $^{32}$P and $^{51}$Cr in man, *International Journal of Nuclear Medicine and Biology*, 6, 17, 1979.
35. Lykken, G.I., A whole body counting technique using ultralow doses of $^{59}$Fe and Zn in absorption and retention studies in humans, *American Journal of Clinical Nutrition*, 37, 652, 1983.
36. Johnson, P.E., Milne, D.B., and Lykken, G.I., Effects of age and sex on copper absorption, biological half-life, and status in humans, *American Journal of Clinical Nutrition*, 56, 917, 1992.
37. King, J.C. and Turnlund, J.R., Human zinc requirements, in *Zinc in Human Biology*, Mills, C.F., Ed., Springer-Verlag, Berlin: Heidelberg, 1989, chap. 21.
38. Aamodt, R.L. et al., Zinc metabolism in humans after oral and intravenous administration of Zn-69m, *American Journal of Clinical Nutrition*, 32, 559, 1979.
39. Arvidsson, B. et al., A radionuclide technique for studies of zinc absorption in man, *International Journal of Nuclear Medicine and Biology*, 5, 104, 1978.
40. Wastney, M.E. et al., Kinetic analysis of zinc metabolism and its regulation in normal humans, *American Journal of Physiology*, 251, R398, 1986.
41. Payton, K.B. et al., Technique for determination of human zinc absorption from measurements of radioactivity in a fecal sample or the body, *Gastroenterology*, 83, 1264, 1982.
42. Flanagan, P.R. et al., Dual-isotope method for determination of human zinc absorption: the use of a test meal of turkey meat, *Journal of Nutrition*, 115, 111, 1985.
43. English, J.L. et al., Use of a dual isotope technique to measure zinc absorption, *FASEB Journal*,, 3, 4954 (abs.), 1989.
44. Sian, L. et al., Influence of a meal and incremental doses of zinc on changes in zinc absorption, *American Journal of Clinical Nutrition*, 58, 533, 1993.
45. Fairweather-Tait, S.J. et al., Zinc absorption in adult men from a chicken sandwich made with white or wholemeal bread, measured by a double-label stable-isotope technique, *British Journal of Nutrition*, 67, 411, 1992.
46. Knudsen, P.E. et al., Zinc absorption estimated by fecal monitoring of zinc stable isotopes validated by comparison with whole-body retention of zinc radioisotopes in humans, *Journal of Nutrition*, 125, 1274, 1995.
47. Schuette, S.A. et al., Dysprosium as a non-absorbable marker for studies of mineral absorption with stable isotope tracers in human subjects, *Journal of American College of Nutrition*, 12, 307, 1993.
48. Friel, J.K. et al., The analysis of stable isotopes in urine to determine the fractional absorption of zinc, *American Journal of Clinical Nutrition*, 55, 473, 1992.
49. Eli Lilly and Co, Indianapolis, IN, U.S., Use of the zinc tolerance test and 24-hour zinc content to assess zinc absorption, *Journal of American College of Nutrition*, 15, 79, 1996.
50. Solomons, N.W., Biological availability of zinc in humans, *American Journal of Clinical Nutrition*, 35, 1048, 1982.

51. Valberg, L.S. et al., Does the oral zinc tolerance test measure zinc absorption?, *American Journal of Clinical Nutrition*, 41, 37, 1985.
52. Sandström, B., Bioavailability of zinc, *European Journal of Clinical Nutrition*, 51 (Suppl. 1), S17, 1997.
53. Molokhia, M. et al., A simple method for measuring zinc absorption in man using a short-lived isotope ($^{69m}Zn$), *American Journal of Clinical Nutrition*, 33, 881, 1980.
54. Scott, K.C. and Turnlund, J.R., A compartmental model of zinc metabolism in adult men used to study effects of three levels of dietary copper, *American Journal of Physiology*, 267, E165, 1994.
55. Barrett, P.H.R. et al., SAAMII: Simulation, analysis and modelling software for tracer and pharmacokinetic studies, *Metabolism*, 47, 484, 1998.
56. Foster, D.M. et al., Zinc metabolism in humans: a kinetic model, *American Journal of Physiology*, 237, R340, 1979.
57. Lowe, N.M. et al., A compartmental model of zinc metabolism in healthy women using oral and intravenous stable isotope tracers, *American Journal of Clinical Nutrition*, 65, 1810, 1997.
58. Bo-Linn, G.W. et al., An evaluation of the importance of gastric acid secretion in the absorption of dietary calcium, *Journal of Clinical Investigation*, 73, 640, 1984.
59. Ramirez, J.A. et al., The absorption of dietary phosphorous and calcium in hemodialysis patients, *Kidney International*, 30, 753, 1986.
60. Fine, K.D. et al., Intestinal absorption of magnesium from food and supplements, *Journal of Clinical Investigation*, 88, 396, 1991.
61. Zheng, J.J. et al., Measurement of zinc bioavailability from beef and a ready-to-eat high-fiber breakfast cereal in humans: application of a whole-gut lavage technique, *American Journal of Clinical Nutrition*, 58, 902, 1993.
62. Wood, R.J. and Zheng, J.J., High dietary calcium intakes reduce zinc absorption and balance in humans, *American Journal of Clinical Nutrition*, 65, 1803, 1997.
63. Raffaniello, R.D. et al., Distinct mechanisms of zinc uptake at the apical and basolateral membranes of Caco-2 cells, *Journal of Cell Physiology*, 152, 356, 1992.
64. Reeves, P. G., Briske-Anderson, M., and Johnson, L., Physiologic concentrations of zinc affect the kinetics of copper uptake and transport in the human intestinal cell model, Caco-2, *Journal of Nutrition*, 128, 1794, 1998.
65. Hansen, M., Sandström, B., and Lonnerdal, B., The effect of casein phosphopeptides on zinc and calcium absorption from high phytate infant diets assessed in rat pups and Caco-2 cells, *Pediatric Research*, 40, 547, 1996.
66. International Programme on Chemical Safety, Environmental Health Criteria 58: Selenium. World Health Organization, Geneva; 1987.
67. Stewart, R.D.H. et al., Quantitative selenium metabolism in normal New Zealand women, *British Journal of Nutrition*, 40, 45, 1978.
68. Bopp, B.A., Sonders, R.C., and Kesterson, J.W., Metabolic fate of selected selenium compounds in laboratory animals and man, *Drug Metabolism Reviews*, 13, 271, 1982.
69. Mutanen, M., Bioavailability of selenium, *Annals in Clinical Research*, 18, 48, 1986.
70. Diplock, A. T., Trace elements in human health with special reference to selenium, *American Journal of Clinical Nutrition*, 45, 1313, 1987.
71. Oldfield, J. E., Selenium in fertilizers, *Bulletin of Selenium-Tellurium Development Association*, Grimbergen, Belgium: Information Center, 1992.

72. Wolffram, S., Mechanismen der intestinalen absorption von selen, *Medizinische klinik*, 90, 1, 1995.
73. Thomson, C.D. and Stewart, R.D.H., The metabolism of [$^{74}$Se]selenite in young women, *British Journal of Nutrition*, 32, 47, 1974.
74. Swanson, C.A. et al., Human [$^{74}$Se]selenomethionine metabolism: a kinetic model, *American Journal of Clinical Nutrition*, 54, 917, 1991.
75. Young, V.R., Nahapetian, A., and Janghorbani, M., Selenium bioavailability with reference to human nutrition, *American Journal of Clinical Nutrition*, 35, 1076, 1982.
76. Moser-Veillon, P.B. et al., Utilization of two different chemical forms of selenium during lactation using stable isotope tracers: an example of speciation in nutrition, *Analyst*, 117, 559, 1992.
77. McConnell, K.P. and Cho, G.J., Transmucosal movement of selenium, *American Journal of Physiology*, 208, 1191, 1965.
78. Barbezat, G.O. et al., Selenium in*: Absorption and Malabsorption of Mineral Nutrients*, Solomons, N. W. and Rosenberg, I. H., Eds., Alan R. Liss, New York, 1984, 213.
79. Sunde, R.A., The biochemistry of selenoproteins, *Journal of American Oil Chemists Society*, 61, 1891, 1984.
80. Patterson, B.H., Levander, O.A., and Helzlsouer, K., Human selenite metabolism: a kinetic model, *American Journal of Physiology* 257 (*Regulatory Integrative Comp. Physiol.* 26), R556, 1989.
81. Atherton, C. et al., Absorption of selenium from biosynthetically labelled foods in humans, in *Trace Elements in Man and Animals — TEMA 10*, Favier, A., Anderson, R.A., and Roussel, A.M., Eds., Plenum, New York, (in press).
82. Linder, M.C., *Nutritional Biochemistry and Metabolism*, Elsevier, New York, 1988, 177.
83. Martin, R.F., Janghorbani, M., and Young, V.R., Kinetics of a single administration of $^{74}$Se-selenite by oral and intravenous routes in adult humans, *Journal of Parenteral and Enteral Nutrition*, 12, 351, 1988.
84. Veillon, C. et al., Selenium utilisation in humans — a long term, self labelling experiment with stable isotopes, *American Journal of Clinical Nutrition*, 52, 155, 1990.
85. Finley, J.W. et al., Selenium supplementation affects the retention of stable isotopes of selenium in human subjects consuming diets low in selenium, *British Journal of Nutrition*, 82, 57, 1999.
86. Griffiths, N.M., Stewart, R.D.H., and Robinson, M.F., The metabolism of [$^{75}$Se]selenomethionine in four women, *British Journal of Nutrition*, 35, 373, 1976.
87. Thomson, C.D. and Robinson, M.F., Urinary and fecal excretions and absorption of a large supplement of selenium: superiority of selenate over selenite, *American Journal of Clinical Nutrition*, 44, 659, 1986.
88. Yergy, A.L., Viera, N.E., and Hansen, J.W., Direct measurement dietary fractional absorption using calcium isotopic tracers, *Biomedical Environmental Mass Spectrometry*, 14, 603, 1987.

# 5

# *Kinetic Studies of Whole-body Trace-element Metabolism*

**Nicola M. Lowe and Malcolm J. Jackson**

**CONTENTS**


5.1 Introduction ... 81
5.2 General Considerations in Study Design ... 82
 5.2.1 Isotope Dose ... 82
 5.2.2 Sampling Strategy ... 82
 5.2.3 Free-Living or Metabolic Unit ... 83
5.3 Compartmental Modelling ... 83
 5.3.1 General Assumptions ... 84
5.4 Specific Examples of Isotope Turnover Studies ... 85
 5.4.1 Zinc ... 85
 5.4.2 Copper ... 86
 5.4.3 Selenium ... 88
5.5 Conclusion ... 89
References ... 90


## 5.1 Introduction

The first report of the use of stable isotopes in the study of the metabolism of essential trace minerals was in the early 1960s, but most of the work in this field has been done in the last 15 years, in parallel with the advances in the technology required to measure stable isotope enrichment in biological tissues.[1] In addition, the availability of ICP-MS instruments has increased dramatically in recent years. For example, information from VG (TJA Solutions, Winsford, Cheshire), one of the leading manufacturers of ICP-MS in the U.K., revealed that in 1987 there were ten VG ICP-MS instruments in use in the U.K. Now, in the year 2000, there are 72. These factors, coupled with the

0-8493-8730-2/00/$0.00+$.50

dramatic surge in development of computer software for the mathematical modelling of biological systems, has fueled studies using stable isotopes to study whole-body metabolism of trace minerals.

In contrast to radioisotopes in which organs can be counted externally (as discussed in Chapter 2), stable-isotope studies are generally limited to the sampling of blood, urine, and fecal material. Nevertheless, a great deal of information about the system can be determined from enrichment data if the sampling regimen is designed so that sufficient samples are taken to allow good temporal resolution of the isotope kinetics over the time period of interest. These kinds of stable-isotope studies have provided important data on absorption and bioavailability of trace minerals (already discussed in detail in Chapter 4). Stable-isotope studies, coupled with compartmental modelling techniques, have also been used to study whole-body, trace-element kinetics and turnover. These studies have provided unique insights into underlying metabolic processes and control mechanisms such as the effects of aging, pregnancy, drugs and diseases, determining nutrient requirements, and establishing optimal dietary intake range.[2–5] It is the use of stable isotopes to study whole-body aspects of trace-mineral metabolism that is the subject of this chapter.

## 5.2 General Considerations in Study Design

There are a number of factors to consider when designing a stable-isotope study. Briefly, these include.

### 5.2.1 Isotope Dose

An estimate can be made of the dose (tracer) required to give an enrichment that can be measured above the native element (tracee) with sufficient precision and accuracy for the time length required. This estimate is based on information such as the initial volume of distribution, the percent enrichment of the tracer administered and its natural abundance, the concentration of tracee in the tissues sampled, and the rate of clearance of the tracer from the tissues sampled. Ideally, the tracer dose must not perturb the mass or kinetics of the system (see Chapter 1).

### 5.2.2 Sampling Strategy

Often a large number of samples is required, particularly in the early time-period after the stable isotope dose, in order to define the rapid changes in isotope enrichment as the isotope is redistributed within the body. Sample preparation is often labor intensive and exacting (Chapter 1). Therefore, careful thought about the timing of blood samples to obtain maximum information with the smallest number of samples is required in order to minimize resources spent on sample preparation and also to remain within the ethical requirements for blood sampling.

### 5.2.3 Free-Living or Metabolic Unit

The number and frequency of sample collections may determine whether the study can be conducted on free-living subjects or those who need to be confined to a metabolic unit. Frequent blood sampling and complete urine or fecal collections are easier if the subjects are confined. In addition, information on the dietary intake of the mineral is often useful for the modelling process, if not essential for the investigation itself. Control and measurement of mineral intake are more reliable if done "in house". This increase in reliability of data needs to be balanced against the disadvantages of metabolic unit confinement: subject recruitment is more difficult for confined studies, the subject dropout rate is higher, and the cost of staffing and running a metabolic unit are considerable.

## 5.3 Compartmental Modelling

Once the samples have been collected and analyzed, the tracer and tracee data can be analyzed together using a mathematical model. One of the difficulties in working with stable isotope tracers, particularly if more than one enriched stable isotope is administered, is that the enriched isotope doses also contain, in small proportions, the other naturally occurring isotopes of that element. The calculations to correct the measured isotope ratios to tracer-tracee ratio have been described in detail[6–8] and are discussed in Chapter 3.

Software tools for modelling are readily available and a lot of work has gone into making them more user friendly. Some of the packages currently available are shown in Table 5.1.

A mathematical model is a hypothesis of how the system works. A compartmental model is composed of a series of compartments representing pools of the element of interest. All the atoms in the pool behave in a kinetically identical way, but may not necessarily be confined to the same physiological location. For example, a compartment may be made of an element from the liver and the red blood cells, which both behave in a kinetically identical manner. Diagramatically, the compartments are connected by arrows representing the movement of the element between the compartments (Figure 5.1). Isotope enrichment from urine, feces, blood, diet intake, and any other independent measures of absorption, along with knowledge of physiology, can be entered into the modelling software. Once the model solution fits observed data and

**TABLE 5.1**

Software Packages Currently Available for Compartmental Modelling

| | |
|---|---|
| SAAM/CONSAM | NIH, Washington, D.C., U.S.A.[9,10] |
| WinSAAM | Laboratory of Experimental and Computational Biology, NIH, Washington, D.C., U.S.A.[11] |
| SAAM II | Resource Facility for Kinetic Analysis, University of Washington, Seattle, U.S.A.[12] |
| STELLA | High Performance Systems, Inc., Hanover, NH, U.S.A.[13] |
| Mlab | Civilised Software, Inc., Bethesda, MD, U.S.A.[14] |
| ACSL | Mitchell & Gauthier Associates, Concord, MA, U.S.A.[15] |
| Scientist | MicroMath Scientific Software, Salt Lake City, UT, U.S.A.[16] |
| Berkeley Madonna | University of California, Berkeley, U.S.A. |

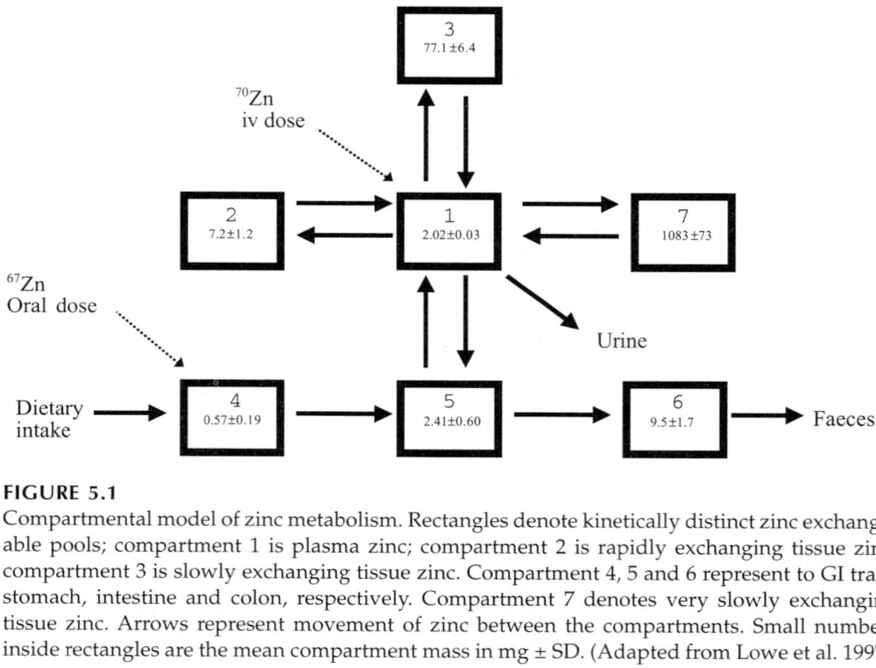


**FIGURE 5.1**

Compartmental model of zinc metabolism. Rectangles denote kinetically distinct zinc exchangeable pools; compartment 1 is plasma zinc; compartment 2 is rapidly exchanging tissue zinc; compartment 3 is slowly exchanging tissue zinc. Compartment 4, 5 and 6 represent to GI tract, stomach, intestine and colon, respectively. Compartment 7 denotes very slowly exchanging tissue zinc. Arrows represent movement of zinc between the compartments. Small numbers inside rectangles are the mean compartment mass in mg ± SD. (Adapted from Lowe et al. 1997.)

is consistent with other known information about a system, the model can be used to calculate parameter values for the system.

### 5.3.1 General Assumptions

Some general assumptions used in constructing a compartmental model include:

- The tracer behaves in the same way as the tracee.
- An i.v. dose behaves the same as an orally administered dose.
- The tracer dose does not perturb the mass or kinetics of the system.

- The tracee is in steady state, i.e., the amount of it in the system is unchanged in the time-frame of the experiment. This is true for most published models of trace-mineral metabolism, but is not a condition for the use of compartmental modelling.

Previously published models provide a useful resource for designing new studies or as a starting point for modelling acquired data. A library for mathematical models of biological systems is currently being compiled and is available on the internet at http://biomodel.georgetown.edu/model/[17]

## 5.4 Specific Examples of Isotope Turnover Studies

### 5.4.1 Zinc

The most detailed models were developed using radioisotopes.[18,19] These multi-compartment models were compiled using external counting of whole-body, liver and thigh regions, as well as sampling of blood. The modelling software SAAM/CONSAM was used to create the model. Using this model, Wastney[17] and co-workers identified five sites for the regulation of zinc homeostasis in humans when intake was varied.[17] The sites include gastrointestinal zinc absorption, urinary zinc excretion, erythrocyte exchange of zinc, muscle zinc release, and secretion of zinc into the gut.

Of the trace elements, zinc is the one that has been most studied by stable-isotope techniques, partly because there are five naturally occurring stable isotopes, two of which ($Zn^{70}$ and $Zn^{67}$) have very low natural abundancies (0.62 and 4.1%, respectively). Jackson and co-workers were one of the first groups to use stable isotopes to study whole-body zinc turnover in humans.[20] They studied a group of undernourished, slum-dwelling, lactating women in Manaus, Brazil using $^{67}Zn$. This population was known to have a marginally deficient dietary zinc intake of 6 mg per day. The purpose of the study was to determine the effect of a chronically low zinc diet on the size and turnover rate of the exchangeable zinc-metabolic pool. The study showed that the exchangeable pool size, studied by measuring isotope disappearance over nine days using a single compartment model (simple isotope dilution), was lower than comparable data from zinc replete subjects. There did not appear to be any significant changes, however, in the turnover rate of this pool.

Since this early work in humans took place, much more complex models of zinc metabolism have been developed. These models are, by necessity, less complex than those derived from radioisotope data, due to the reduced amount of data that can be collected from stable-isotope studies. However, they are still useful tools in the study of zinc metabolism. A multi-compartment model of zinc metabolism had been developed using orally ($^{67}Zn$) and intravenously ($^{70}Zn$) administered stable isotopes (Figure 5.1).[21] The study was conducted in a group of healthy young women who consumed a constant diet containing 7.0 mg Zn/day. The zinc isotopes were administered simultaneously. Multiple plasma and 24-hour urine samples were collected over the following 7 days, with complete fecal collections for 11 days. Isotope enrichment data, along with naturally occurring zinc levels, were analyzed using SAAM/CONSAM. The purpose was to develop a model with the least number of compartments required to account for the dynamic properties of the data, while remaining consistent with known physiology. The resulting model enabled calculation of fractional absorption of zinc from the gastrointestinal tract, the rates of endogenous secretion and excretion, the fractional turnover rate of the plasma pool, and the sizes and fractional turnover rates of extravascular pools that exchange with plasma zinc.

This model was applied to investigate the zinc homeostatic mechanism in a group of men participating in a zinc depletion study.[22] The study was performed in a metabolic unit and the subjects consumed a semi-synthetic formula diet containing 12.2 mg of zinc during the 16-day baseline period and 0.23 mg of zinc per day during the 41-day depletion period. Stable isotopes were administered intravenously at the mid-point of the baseline period and at the end of the depletion period. Isotope enrichment of the plasma, urine, and feces from both metabolic periods was analyzed concurrently using the SAAM II. Under these conditions of severe zinc deficiency, the plasma zinc mass decreased to 35% of its baseline value, from an average of 3.6 mg to 1.2 mg. The model suggested fractional zinc absorption increased to close to unity and the excretion of zinc into the feces and urine decreased by 91 and 99%, respectively, to establish a new, near-steady state at depletion. In order to determine how these changes occurred, a dynamic model of zinc mass movement was formulated, based on the average values of the rate constants from the tracer model at baseline. Changes in the rate constants from baseline to depletion were tested individually to determine their effect on plasma mass. The only rate constant that could explain the dramatic changes in the plasma zinc mass during depletion was the fractional rate of transfer of zinc from compartment seven (very slowly turning over tissues) to compartment one (plasma zinc). This rate constant changed from a baseline value of 0.015/day to 0.006/day at depletion. This slowly turning-over zinc pool represents more than 90% of the total body zinc mass and is composed primarily of muscle zinc. The model therefore suggests that the zinc pool is very sensitive to acute dietary zinc restriction so that, when zinc intake is low, the amount of zinc from these tissues returning to the plasma pool is decreased significantly.

Simpler models derived from stable-isotope enrichment data have also been used to study zinc status and are discussed in more detail in Chapter 7.[23,24]

### 5.4.2 Copper

There are only two stable isotopes of copper, $^{63}Cu$ and $^{65}Cu$, with natural abundancies of 69% and 31%, respectively. Enriched $^{65}Cu$ isotope can be used as the "tracer"; however, relatively large doses are required in order to achieve a detectable enrichment of plasma above natural levels.

Copper turnover studies using enriched $^{65}Cu$ have been used as a tool in the diagnosis of Wilson's disease and have proved particularly valuable in identifying non-symptomatic heterozygotes for this condition.[25,26] Wilson's disease is a genetic disorder of a copper transporting ATPase on chromasome 13.[27–29] This affects the excretion of copper in the bile and the incorporation of copper into ceruloplasmin, the main copper-binding protein in the plasma.

Lyon and coworkers gave an oral dose of 3 mg $^{65}Cu$ (99% enriched); blood samples were taken at 1, 2, 6, 24, 48, and 72 hours.[25] In healthy patients, when $^{65}Cu$ is absorbed from the gastrointestinal tract, the resulting plasma enrichment profile shows an initial enrichment peak at 1 to 2 hours, followed by a minimum at 6 hours. Plasma enrichment then rises slowly again over a period of several days, as the isotope is incorporated into ceruloplasmin. In Wilson's disease, this secondary rise is missing. The heterozygotes, who carry the WD gene but do not suffer from the disease, have profiles that range from normal to those with no secondary rise (also reported by Merli[26] et al., 1998). In addition, the biological half-time for clearance of $^{65}Cu$ from the plasma pool was increased in the heterozygotes compared to the control subjects (43 days vs 18.5 days).

In order to study copper homeostasis, Scott and Turnlund developed a multi-compartment model of copper metabolism from plasma $^{65}Cu$ enrichment data collected during a 90-day study.[30] $^{65}Cu$ was used as a tracer of copper metabolism in adult men at low, adequate, and high levels of dietary copper. The study was divided into three metabolic periods (MP): MP1 dietary copper levels were adequate (1.68 mg per day for 28 days); MP2 dietary copper levels were low (0.79 mg per day of 42 days), and MP3 dietary copper levels were high (7.53 mg per day for 24 days). Subjects were housed in a metabolic unit, and $^{65}Cu$ was administered intravenously, once during each metabolic period. Blood was sampled at various time intervals up to 2 days postinfusion. In order to measure fractional copper absorption, $^{65}Cu$ was administered orally once during MP1 and MP2 and twice during M3.[31] Plasma isotope ratios were measured using TIMS and the data modelled using CONSAM.

The model that provided the best fit to the observed data consists of five compartments, two delay components, and two routes of excretion. Based on known physiology of copper metabolism, these were interpreted to represent two plasma compartments (one containing ceruloplasmin-bound copper and one non-ceruloplasmin-bound copper), two liver and one other tissue compartments and urinary and fecal excretion. The model was fit individually to data from each metabolic period. The parameter values at different dietary levels were compared to identify those involved in the homeostatic response. The model demonstrated that dietary copper level influenced the rate of transfer for two pathways: the transfer from the second liver compartment to the second plasma compartment was lower by 30% when dietary copper was low. The fractional rate of transfer of copper from ceruloplasmin-bound plasma compartment to other tissues varied with level of copper intake, transfer being lowest (0.39 per hour) during the low copper intake period and highest (0.78 per hour) during the high copper intake period.

### 5.4.3 Selenium

Plasma selenium kinetics are complex because of the large number of different forms of selenium in the plasma.[32] These include

- Glutathione peroxidase
- Selenoprotein P
- Seleno-aminoacid (e.g., selenomethionine)
- Dimethyl selenide

The above comprise the functionally important selenium, derived from selenite. There is another major pool of selenium in the form of selenomethionine which has no known biochemical function and is thought to be a storage compartment for Se.

A compartmental model describing the relationship between these selenium species has been developed by Patterson and coworkers.[33,34] Six subjects were given a 200 μg oral dose of the stable isotope $^{74}$Se as either selenite or selenomethionine. Plasma samples were collected 30 minutes after dose administration, then hourly for the next 8 hours, daily for 6 days, and then weekly for 2 to 3 weeks. Feces were collected daily for 12 days to characterize absorption. Twenty-four-hour urine collections were made for 11 days. Using the modelling program SAAM/CONSAM, a detailed model of selenium metabolism has been described.[34] In addition to gastrointestinal compartments, peripheral tissue, and hepatopancreatic compartments, the model identifies four kinetically distinct plasma pools representing chemically different forms of selenium. The authors suggest that, since each of the different forms of plasma selenium may perform different roles in terms of nutritional requirements, toxicity and cancer prevention, this model provides a starting point for further studies to identify the forms of selenium with the greatest beneficial potential, and to examine their kinetics in more detail.

A reliable means of assessing selenium status is important in order to establish a rational nutrition policy with respect to dietary selenium intakes. Alternative biochemical methods for evaluation of selenium status in humans include the measurement of the activity of selenium-dependent enzymes, or the concentration of selenium in accessible body fluids and tissues (see Chapter 9). Selenium concentration in plasma, or other cellular elements of blood, do not respond to very high or very low levels of selenium intake.[35]

An alternative approach is to measure the size of the body selenite pool using isotope dilution.[32] This technique depends upon the dilution of a single oral or intravenous dose of $^{74}$Se in the selenite-exchangeable metabolic pool (Se-EMP). The method has a number of assumptions which include the following

- There are two biochemically distinct pools of selenium in the body. One consists of selenocompounds that can be derived from selenite. The second pool consists of selenomethionine containing proteins.
- Labelling of the Se-EMP results in progressive labelling of the components of Se-EMP, but not from selenomethionine.
- Isotope ratios in urine and plasma ultrafiltrate will reach identical value; therefore, urine can be used instead of plasma.

100 μg of enriched $^{74}$Se as selenite was infused in 500 ml of isotonic saline over a period of 4 hours. A complete 24-hour baseline urine collection was made the day before the isotope infusion, and for 14 successive days commencing 24 hours following isotope infusion. The size of the Se-EMP was determined from isotope enrichment in the urine. Using this technique, Janghorbani et al. demonstrated that Se-EMP decreased by 16% in response to a restriction of dietary selenium intake.[32] The main advantage of this approach is that the measurement reflects the size of the pool present in the whole body in contrast to plasma concentrations of selenium which may have little relationship to intracellular concentrations of the trace element.

The kinetics of two rapidly turning-over selenite pools have also been used to study the effects of age and institutionalization on selenium metabolism and status. Using a simple two-compartment model, in 1997, Ducros at al. analyzed the plasma disappearance kinetics of an intravenous dose of 100 μg $^{74}$Se in three groups of women: a group of institutionalized elderly, a group of free-living elderly, and a group of young adults.[2] Plasma isotope enrichment was measured for up to 6 months and the resulting plot of isotope enrichment vs time was found to follow a double exponential equation describing two compartment kinetics.

The model enabled the size of the two exchangeable selenium pools to be determined, along with their half-lives, and the rate constants describing the movement of Se between the two compartments and out of the system (elimination). The study demonstrated that there were significant differences in selenium kinetics in the institutionalized elderly subjects compared to the young adult group: the size of the first pool (Qa) was significantly reduced in the institutionalized elderly group compared to the other two groups, and the turnover rate of both pools was higher in this group than the young adult group. In addition, the elimination rate was lower in the institutionalized group, indicating greater retention of $^{74}$Se.

## 5.5 Conclusion

Stable-isotope studies are undoubtedly expensive and labor-intensive, but have the advantage that, when coupled with mathematical modelling, they can provide a great deal of novel and unique information that would otherwise be unobtainable (for example, from tissues that cannot be sampled directly). This chapter has reviewed some of the ways in which this approach has furthered our knowledge of trace-element metabolism and homeostasis. Now that models of trace-mineral metabolism developed using data from healthy human subjects are becoming accepted and established, there is a great deal of potential for future work to use these to study effects of disease and pharmaceuticals on mineral metabolism.

## References

1. Turnlund, J.R., The use of stable isotopes in mineral nutrition research, *J. Nutr.*, 119, 7, 1989.
2. Ducros, V. et al., The sizes of the exchangeable pools of selenium in elderly women and their relation to institutionalization, *Br. J. Nutr.*, 78(3), 379, 1997.
3. O'Brien, K.O., Regulation of mineral metabolism from fetus to infant: metabolic studies, *Acta. Pediatr. Suppl.*, 88(433), 88, 1999.
4. Fung, E.B. et al., Zinc absorption in women during pregnancy and lactation: a longitudinal study, *Am. J. Clin. Nutr.*, 66(1), 80, 1997.
5. Lowe, N.M. et al., Studies of human zinc kinetics using the stable isotope $^{70}Zn$, *Clin. Sci.*, 84, 113, 1993.
6. Buckley, W.T., The use of stable isotopes in studies of mineral metabolism, *Proceedings Nutr. Soc.*, 47, 407, 1988.
7. Buckley, W.T., Huchin S.N., and Eigendorf G.K., Calculation of stable isotope enrichment for tracer kinetic procedures, *Biomed. Mass Spectrom.*, 12, 1, 1985.
8. Cobelli, C., Toffolo, G., and Foster, D.M., Tracer-to-tracee ratio for analysis of stable isotope tracer data: link with radioactive kinetic formulism, *Am. J. Physiol.*, 262, E968, 1992.
9. Berman, M. and Weiss, M.F., SAAM Manual, Washington DC: U.S. Printing Office, DHEW Publication No. (NIH) 78–180, 1978.
10. Berman M. et al., CONSAM User's Guide, Washington, DC: U.S. Govt. Printing Office, 3279, 421, 1983.
11. Greif, P. et al., Balancing needs, efficiency, and functionality in the provision of modelling software: a perspective of the NIH WinSAAM project, *Mathematical Modelling in Experimental Nutrition*, Plenum Press, New York, 1998, chap. 1.
12. SAAM II: A Program for Kinetic Modelling, Resource Facility for Kinetic Analysis, University of Washington, Seattle, U.S.A., 1997.
13. Stella II: An Introduction to Systems Thinking. High Performance Systems, Inc.: Hanover, NH, U.S.A., 1992.
14. Mlab: A Mathematical Laboratory, Civilised Software, Inc.: Bethesda, MD, U.S.A., 1996.
15. ASCL: Advanced Continuous Simulation Language, Version 10.2, Mitchell and Gauthier Associates: Concord, MA, U.S.A., 1992.
16. Scientist: For Experimental Data Fitting, Version 2.0, MicroMath Scientific Software: Salt Lake City, UT, U.S.A., 1995.

17. Wastney, M.E. et al, *Mathematical Modelling in Experimental Nutrition*, Plenum Press, New York, 1998, chap. 8.
18. Foster D.M. et al., Zinc metabolism in humans; a kinetic model, *Am. J. Physiol.*, 237(5), R340, 1979.
19. Wastney, M.E. et al., Kinetic analysis of zinc metabolism and its regulation in normal humans, *Am. J. Physiol.*, 251, R398, 1986.
20. Jackson, M.J. et al., Stable isotope metabolic studies of zinc nutrition in slum-dwelling lactating women in the Amazon valley, *Br. J. Nutr.*, 59, 193, 1988.
21. Lowe N.M. et al., A compartmental model of zinc metabolism in healthy women, using oral and intravenous stable isotopes, *Am. J. Clin. Nutr.*, 65, 1810, 1997.
22. King, J.C. et al., Effect of acute zinc depletion in men on zinc homeostasis and plasma zinc kinetics, *Am. J. Clin. Nutr.*, in press.
23. Miller, L.V. et al., Size of the zinc pools that exchange rapidly with plasma zinc in humans: alternative techniques for measuring and relation to dietary zinc intake *J. Nutr.*, 124, 268, 1994.
24. Fairweather-Tait, S.J. et al., The measurement of exchangeable pools of zinc using the stable isotope $^{70}Zn$, *Br. J. Nutr.*, 70, 221, 1993.
25. Lyon, T.D. et al., Use of a stable copper isotope ($^{65}Cu$) in the differential diagnosis of Wilson's disease, *Clin. Sci.*, 88(6), 727, 1995.
26. Merli M. et al., Use of the stable isotope $^{65}Cu$ test for the screening of Wilson's disease in a family with two affected members, *Ital. J. Hepatol.*, 30(3), 270, 1998.
27. Bull, P.C. et al., The Wilson's disease gene is a putative copper transporting P-type ATPase similar to the Menkes gene, *Nature (Genetics)*, 5, 327, 1993.
28. Tanzi, R.E. et al., The Wilson's disease gene is a copper transporting ATPase with homology to the Menkes disease gene, *Nature (Genetics)*, 5, 344, 1993.
29. Petrukin, K. et al., Mapping, cloning and genetic characterisation of the region containing the Wilson's disease gene, *Nature (Genetics)*, 5, 338, 1993.
30. Scott, K.C. and Turnlund J.R., Compartmental model of copper metabolism in adult men, *J. Nutr. Biochem.*, 5, 342, 1994.
31. Turnlund, J.R. et al., Copper absorption and retention in young men at three levels of dietary copper by use of the stable isotope $^{65}Cu$, *Am. J. Clin. Nutr.*, 49, 870, 1989.
32. Janghorbani, M. et al., The selenite-exchangeable metabolic pool in humans: a new concept for the assessment of selenium status, *Am. J. Clin. Nutr.*, 51, 670, 1990.
33. Patterson, B.H. and Zech, L.A., Development of a model for selenite metabolism in humans, *J. Nutr.*, 122, 709, 1992.
34. Patterson, B.H. et al., Human selenite metabolism: a kinetic model, *Am. J. Physiol.*, 257, R556, 1989.
35. Robinson, M.F., *Clinical, Biochemical, and Nutritional Aspects of Trace Elements*, Alan R. Liss Inc., New York, 1982, 325.

# 6

# *Stable-isotope Methods for the Investigation of Iron Metabolism in Man*

**Morteza Janghorbani**

**CONTENTS**

6.1 Introduction ........ 93
6.2 Iron Metabolism in Relation to the Design of Stable-isotope Protocols ... 94
6.3 Feasibility Issues ........ 95
6.4 Analytical Methods ........ 99
6.4.1 Neutron Activation Analysis ........ 99
6.4.2 Mass Spectrometry ........ 100
6.4.3 Summary of Current Analytical Capabilities ........ 101
6.5 Selected Applications ........ 102
6.5.1 Relationship between Mucosal Absorption and Hemoglobin Incorporation of Dietary Iron ........ 102
6.5.2 Issues of Dietary Availability of Iron ........ 103
6.6 Conclusion ........ 104
References ........ 105

## 6.1 Introduction

Investigations of certain important aspects of iron metabolism require use of isotopic tracers. Traditionally, two radioactive isotopes of iron, $^{55}Fe$ and $^{59}Fe$, have been used for this purpose. However, concerns about radiation exposure have limited the use of these isotopes for a wide range of important applications: dietary availability of various forms of iron (especially in infants, children, and women of child-bearing status), regulation of iron metabolism in iron overload disorders, etc. This limitation, coupled with recent advances in the analytical methods for accurate measurement of stable isotopes of iron in biological materials, has provided strong impetus for

0-8493-8730-2/00/$0.00+$.50

**TABLE 6.1**

Stable Isotope Composition of Iron

| Stable isotope | Natural Abundance (atom, %) | Typical Available Enrichment (atom, %) |
|---|---|---|
| $^{54}Fe$ | 5.8 | 0.39 |
| $^{56}Fe$ | 91.7 | 13.78 |
| $^{57}Fe$ | 2.14 | 1.25 |
| $^{58}Fe$ | 0.31 | 84.58 |

applications of stable isotopes of iron in investigations of iron metabolism. This chapter provides a brief review of the current state of the methodology of stable isotopes as they apply specifically to investigations of iron metabolism in man and highlights some of its applications.

Iron consists of four stable isotopes with natural abundances as shown in Table 6.1. Isotopically enriched iron is available with enrichment for any of the isotopes, as shown in the example in this table.[1] Meaningful applications of stable isotopes of iron necessitate a sufficient understanding of some important issues that could pose significant limitations for the conduct of investigations:

1. In contrast to radio-iron, stable isotopes of iron are used at substrate levels.
2. Measurement of stable isotopes of iron requires a much higher level of sophistication than measurement of their radioactive counterparts.
3. Use of stable isotopes of iron involves significant expense. Because of the fundamental significance of the first two issues in the proper design of metabolic investigations, and their role in determining the isotopic costs of an experiment, they are discussed here in some detail.

## 6.2 Iron Metabolism in Relation to the Design of Stable-isotope Protocols

Essential for proper design of stable-isotope investigations is an understanding of selected quantitative aspects of iron metabolism: iron balance and ferrokinetics in the plasma and red blood cell (RBC) pools. Data on iron balance and its body distribution in healthy adults with normal iron stores are summarized in Table 6.2. Typical daily intake of iron is 10 to 20 mg. Since obligatory losses are about 0.6 mg for men and 1.2 mg for menstruating women, iron balance requires that only a small fraction of daily intake is absorbed. In a typical, healthy male with normal stores, 2.4 g of the total body iron content

**TABLE 6.2**

Iron Composition of Healthy Adults with Normal Iron Stores

| | |
|---|---|
| Total | 3.5 g (50 mg/kg) in men; 2.1 g (37 mg/kg) in women |
| Iron in circulating hemoglobin | 2.4 g in men; 1.5g in women |
| Plasma iron | 3 mg in men; 2.5 mg in women |
| Daily intake | 10–20 mg/day |
| Daily losses | 0.6 mg/day for men; 1.2 mg/day for menstruating women |
| Daily fecal output | 10–20 mg/day |

of 3.5 g is in circulating hemoglobin, and only a small quantity (3 mg) is present in plasma.

Design of stable-isotope protocols is more complex than radio-iron studies because of the existence of substantial stable-isotope backgrounds in metabolic compartments of interest. For this purpose, steady-state iron balance data, while important, are not sufficient by themselves to permit evaluation of the isotope dosing requirements and prediction of the likely enrichments. This evaluation is especially important if isotopic enrichment in plasma is of interest. Ferrokinetic parameters must be taken into account for the design of such studies. Ferrokinetics may vary markedly in patients with different disorders of iron metabolism compared with healthy subjects with normal iron stores, and these affect both the rate of plasma turnover and hemoglobin incorporation of absorbed stable isotopes.[2] As a result, the extent of isotope enrichment observed in any compartment, and thus the dosing requirements, are directly affected. Because of their importance in determining the extent of isotopic enrichment that can be achieved, selected data on plasma turnover ($t_{1/2}$) and red cell utilization are summarized in Table 6.3.

## 6.3 Feasibility Issues

Stable isotopes of iron can be used to investigate issues of iron absorption, plasma kinetics, and red cell incorporation. However, because of the presence of significant background levels in various metabolic compartments and marked differences in kinetics of isotope turnover in different iron compartments, a brief discussion of important aspects of experimental feasibility is needed.

From a feasibility point of view, the central issue is the achievable degree of isotopic enrichment in the compartment of interest in relation to both isotope cost and scientific validity of the experimental protocol. In general, the most desirable dose is that which provides scientifically valid data at minimum cost. Data illustrating the relationship between required extent of isotope enrichment and resultant cost are summarized in Table 6.4. Let us assume, for instance, that the purpose of the experiment is to investigate incorporation of isotope in circulating red blood cells (RBCs) in an adult male with normal

**TABLE 6.3**

Plasma Turnover and Red Cell Utilization Observed in Selected Disorders of Iron Metabolism

| Disorder | Plasma Turnover ($t_{1/2}$) (minutes) | Red Cell Utilization (%) |
|---|---|---|
| Normal | 86 | 80 |
| Iron deficiency anemia | 20 | 94 |
| Hypoplastic anemia | 310 | 26 |
| Hemochromatosis | 121 | 74 |

iron metabolism and stores. The background levels of $^{58}$Fe and $^{57}$Fe of the circulating RBC-pool in this individual are 7.8 and 52 mg, respectively. If we assume that the isotope measurement method available to us permits measurement precision (percent relative standard deviation of the isotope ratio, %RSD) of 1%, then the minimum amount of enriched isotope required at the time of sampling is 0.78 and 5.2 mg for each isotope, respectively. (A reasonable rule is that isotope enrichment should be ten times the measurement precision.)[3] The cost of the isotope present in the circulating RBC-pool at the time of sampling will be $203 and $82, respectively. If the investigation involves injecting the isotope, the resultant total isotope cost for each subject is then only somewhat higher than this because approximately 80% of injected dose is expected to be incorporated in circulating hemoglobin (Table 6.3). In contrast, if the purpose of the experiment is to investigate hemoglobin incorporation of ingested isotope, then the cost is higher in direct proportion to the fraction of the ingested dose that is absorbed. For absorption of 10%, for example, isotope cost will be $2,030 and $820 for the two isotopes, respectively. In direct contrast, the cost of achieving a similar level of isotope enrichment in plasma will be trivial even if the isotope is to be ingested under the conditions of low absorption (Table 6.4). The example here is important not only from a cost point of view, but also to establish whether a physiologically relevant experiment can be designed. For instance, if 5.2 mg of $^{57}$Fe is to be injected, the design of the experiment must be such that this relatively large amount of Fe, in comparison to circulating plasma Fe of 3 mg, does not disturb the plasma iron pool, thus leading to unphysiologic response.

For experiments involving injection of stable isotopes and measurement of isotope enrichment in circulating RBCs and/or plasma, the issues of feasibility are straightforward as seen from data given in Table 6.4. However, the majority of past and current applications of stable isotopes of iron have focused on issues of gastrointestinal absorption.[4–24] These applications require administration of an appropriate dose of one, or more, of the isotopes orally in a suitable chemical form, followed by either measurement of fecal excretion of the isotope, its incorporation in circulating hemoglobin 10 to 14 days later, or concurrent intravenous administration of a second isotope followed by measurement of isotope enrichment in plasma within a few hours or in circulating RBCs 10–14 days later.[10,11,16–18,20,25] The requirements

**TABLE 6.4**

Data Needed to Establish Feasibility of a Metabolic Study in an Adult Male with Normal Iron Metabolism

| | $^{58}Fe$ | $^{57}Fe$ |
|---|---|---|
| Circ. RBC | 7.8 mg | 52 mg |
| Circ. plasma | 9.7 μg | 65 μg |
| Measurement precision for $^{58}Fe/^{54}Fe$ (or $^{57}Fe/^{54}Fe$) | 1% | 1% |
| Required min. enrichment | 10% | 10% |
| Required min. dose in | | |
| Circ RBC | 780 μg | 5.2 mg |
| Circ. plasma | 0.97 μg | 6.5 μg |
| Cost (US$) for min. dose for precision = 1%[a] | | |
| Circ. RBC | 203 | 82 |
| Circ. plasma | 0.25 | 0.10 |
| Cost (US$) for min. dose for precision = 0.1% | | |
| Circ. RBC | 20 | 8 |
| Circ. plasma | 0.03 | 0.01 |

[a] Based on $213/(mg Fe) for $^{58}Fe$ (82%); $14.5/(mg Fe) for $^{57}Fe$ (92%). Oak Ridge National Laboratory, 1999.

related to achievable isotope enrichment for protocols involving oral administration of stable isotopes vary widely depending on the specific matrix used for monitoring isotope enrichment, but the issues of cost may preclude conduct of some of the protocols.

In protocols involving measurement of fecal excretion of a stable isotope of iron following its ingestion, determining the minimum required dose is straightforward. These protocols usually involve administration of an appropriate form of the isotope as a single dose or several doses spread over one or more days. Thus, the issue of optimizing the required dose only involves determination of the fecal background of the isotope in which it appears. Keep in mind, however, that achieved isotope enrichment in the fecal sample needs to be equal to at least ten times the precision with which the appropriate isotope ratios can be measured. For instance, if the daily intake of iron for the subjects is 10 mg, resulting in a similar daily fecal excretion, then daily fecal background for $^{58}Fe$ and $^{57}Fe$ will be 32 and 218 μg, respectively. Since only a small portion (<50%) of the ingested dose is generally absorbed, the minimum oral dose is approximately 6 and 40 μg for $^{58}Fe$ and $^{57}Fe$, respectively. Therefore, for such protocols, isotope cost is a minor issue.

Protocols that focus on issues of dietary availability of iron and involve measurement of fecal excretion of ingested isotope(s) are likely to provide useful data only if the expected gastrointestinal absorption is relatively high.[3] As fractional gastrointestinal absorption of a dose of the ingested stable isotope is decreased, the error resulting from measurements of intake and fecal excretion becomes more prominent. This is an important design issue and has been discussed in detail previously.[3] In general, for elements such as iron in which gastrointestinal absorption from diet is relatively low, monitoring excretion of the ingested dose does not provide an accurate approach for assessment of absorption. The exception is for situations where the expected gastrointestinal absorption is relatively high, such as iron absorption in iron-deficient patients or absorption of a reference dose.

Thus, a suitable experimental approach for assessment of gastrointestinal absorption of an ingested dose of iron, in most situations involving the issue of dietary availability, is to monitor appearance of the absorbed stable isotope in blood plasma or in circulating RBCs. It is important to understand that the features and limitations of each approach are unique. Data given in Table 6.4 regarding hemoglobin incorporation clearly demonstrate that achieving sufficient degree of isotope enrichment in circulating RBCs, in relation to precision of isotope measurement, is possible only if: 1) gastrointestinal absorption is relatively high; 2) $^{58}Fe$ is used for oral dosing; or 3) the subjects are infants (i.e., hemoglobin pool is small). The data also show the major role that improved measurement precision plays in providing realistic options for practical protocols.

Because of the difficulties inherent in both the fecal isotope balance approach and hemoglobin incorporation as realistic general approaches to the measurement of iron absorption using stable isotopes, examining the features and limitations of monitoring plasma becomes of interest. Data given in Table 6.4 illustrate that sufficient isotopic enrichment can be readily achieved with small quantities of either $^{58}Fe$ or $^{57}Fe$ entering the circulating transferrin pool. This, combined with the reasonably long residence time ($t_{1/2}$) of absorbed stable isotope in plasma (Table 6.3), indicates that plasma may be a suitable compartment for investigations of iron absorption in a wide range of patients if sufficient hemoglobin enrichment cannot be achieved. This is especially important in studies focusing on dietary availability of iron. To illustrate the potential significance of this, consider the amount of $^{58}Fe$ that needs to be ingested by an adult whose iron absorption is only 1%. If plasma samples are obtained over a time period corresponding to two half-lives after absorption of the isotope (approximately 3 hours for a healthy adult; Table 6.3), a 200-μg dose of $^{58}Fe$ will be sufficient to achieve the required plasma enrichment. Thus, plasma sampling may provide unique capabilities for studies that cannot be conducted with either the fecal isotope balance procedure or the hemoglobin incorporation method, due to limitations of iron absorption and large stable isotope background of circulating RBCs. However, this potential major advantage is offset by some potentially serious limitations. First, plasma sampling cannot be used for determination of absolute value of absorption unless, concurrently with ingestion of the isotope, a second isotope is administered intravenously in a form that allows its ready incorporation into circulating plasma transferrin. The ratio of the two isotopes in a sample of plasma obtained some hours after their co-administration then provides absolute value of absorption for the ingested isotope. Because the amount of isotope required for injection is small compared with plasma circulating iron (Table 6.4), disturbance of the transferrin kinetics is not an important limitation. A second approach is to measure absorption relative to that for a reference dose, which is the common practice in studies of iron availability. In this approach, absolute value of absorption of the reference dose can be derived from either fecal isotope balance or hemoglobin incorporation of the isotope from the reference dose, so that this, combined with the relative values of absorption between the test dose and the reference dose obtained from plasma sampling, permits determination of absolute value of absorption for the test dose. Thus, the major limitation of plasma sampling is related to the need for drawing multiple blood samples, which restricts its application, especially in infants and children.

## 6.4 Analytical Methods

Applications of stable isotopes of iron require that suitable analytical methods for precise and accurate measurement be available. During the past three decades, significant advances have occurred in this area, so that, currently, a number of methods are available that can be used for routine applications.[4–6,11,16–20,22,26–30] These methods are based on one of two basic analytical approaches: neutron activation methods of analysis, or methods based on mass spectrometry.[4–6,11,16–22,25–30]

### 6.4.1 Neutron Activation Analysis

Neutron activation analysis (NAA), as applied to the measurement of stable isotopes of iron, is based on capture of a thermal neutron by the nucleus of stable isotopes of iron and subsequent conversion to a radioactive isotope with suitable radioactive emissions. The intensity of the radioactive emissions is proportional to the number of nuclei of the stable isotope undergoing the nuclear reaction. The most suitable source of thermal neutrons for this application is a research reactor with appropriate thermal neutron flux. Of the four stable isotopes of iron, only two ($^{54}Fe$ and $^{58}Fe$) have the required characteristics.[27] Thermal neutron capture by $^{54}Fe$ produces $^{55}Fe$, which emits X-rays (5.9 and 6.59 kev). Measurement of these X-rays requires preparation of a thin target and an appropriate X-ray detection system. On the other hand, capture of thermal neutrons by nuclei of $^{58}Fe$ produces $^{59}Fe$ which emits $\gamma$-rays (1099 kev) and can be measured with standard high-resolution $\gamma$-ray measurement instrumentation. The latter measurement can be carried out without the need for performing any chemistry. A method for simultaneous measurement of both $^{54}Fe$ and $^{58}Fe$ has been published specifically for application to stable isotope studies in humans.[27] In this method, the biological sample (blood, feces) is first dissolved by wet ashing, iron is precipitated, and the precipitate is irradiated in the reactor. The irradiated iron from the sample is precipitated as $Fe(OH)_3$ on a filter paper to provide the necessary thin target. This is then counted simultaneously for both X-rays ($^{55}Fe$) and

$\gamma$-rays using a suitable combination of X-ray and $\gamma$-ray detection.[27] Measurement precision of the ratio $^{54}Fe/^{58}Fe$ for this method for routine analyses has been reported to be in the range of 1 to 5%. The limitations of this method are related to relative lack of availability of neutron activation facilities, needed instrumentation, and radiochemical expertise. When available, this is an excellent method for routine applications as long as its measurement precision is satisfactory, since it permits large number of samples to be processed and requires relatively simple chemical manipulations.

In cases where the instrumentation required for simultaneous measurement of $^{54}Fe$ and $^{58}Fe$ with the technique above is not available, neutron activation analysis of $^{58}Fe$ can be combined with standard methods for measurement of elemental iron (Fe).[11,16] In this approach, overall precision of $^{58}Fe/Fe$ is determined by the appropriate combination of the precision of the two measurements and is typically in the range of 1 to 5%.[16] Assuming that the overall precision is not much worse than that for $^{54}Fe/^{58}Fe$ method (above), this method should be as satisfactory.

### 6.4.2 Mass Spectrometry

Several variations of mass spectrometry (MS) have been developed for measurement of stable isotopes of iron: chelate MS,[26] inductively couple plasma MS (ICP-MS), fast atom bombardment MS (FAB-MS), and thermal ionization MS (TI-MS).[4–6,11,18,22,25–28,30] Each method has unique advantages and limitations. For instance, ICP-MS is a reasonably rapid method suitable for routine analyses of $^{54}Fe$, $^{57}Fe$, and $^{58}Fe$, but not $^{56}Fe$.[28] All MS methods require some degree of chemical separation, the nature and extent of which varies with the specific technique and the matrix of the biological material. For instance, measurement of $^{58}Fe/^{57}Fe$ in blood using ICP-MS can be accomplished with a relatively simple procedure involving digestion of the sample followed by precipitation and redissolution of its iron.[25] In contrast, the same measurement in fecal material requires exhaustive separation of fecal Fe from fecal Ni because $^{58}Ni$ is present in high concentration relative to $^{58}Fe$.[28] A typical procedure for routine measurement of isotope ratios with ICP-MS in biological matrices of relevance is summarized in Figure 6.1.

Analytical instrumentation used for stable isotope investigations should be evaluated in terms of the following criteria: basic capabilities, ease of operation, and overall costs. The criteria for basic capabilities include breadth of applications, overall measurement precision, extent of chemical manipulations required, and sample throughput. Neutron activation analysis is basically limited in its ability to measure a wide range of stable isotopes; for iron, only two stable isotopes can be measured. In contrast, MS methods have the inherent capability for measurement of a wide range of stable isotopes. Examination of the available literature shows that, for routine measurements, analytical precisions are similar — around 0.5 to 5% for all the currently available MS techniques; none can reliably provide measurement precision of

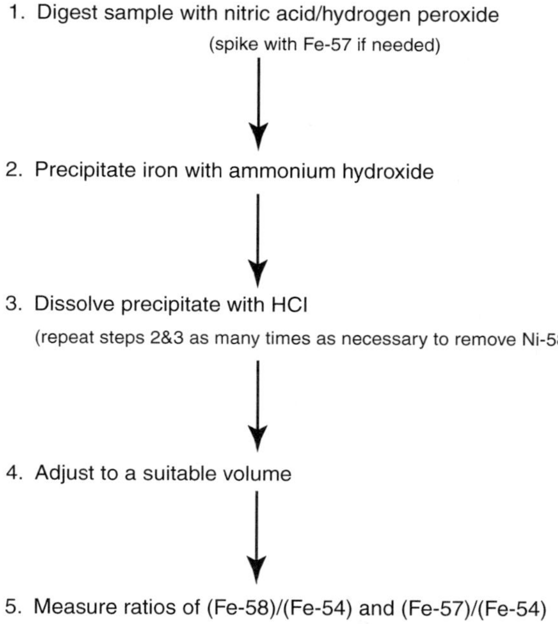


**FIGURE 6.1**
Summary of analytical scheme for measurement of isotope ratios in biological matrices enriched *in vivo* with Fe-58 using inductively coupled plasma mass spectrometry (ICP-MS).

better than 0.5% for routine measurements.[4–6,11,18–22,25–28,30] Sample throughput depends on the required extent of chemical separation and varies considerably among various techniques and for different matrices. While instrument costs may vary significantly, they are all expensive, requiring skilled personnel for their operation. On balance, the current literature indicates that, while any of the MS techniques may present special advantages for particular circumstances, none is superior in general terms.

### 6.4.3 Summary of Current Analytical Capabilities

Advances made during the past 30 years toward development of suitable analytical methods for accurate measurement of stable isotopes in general, and for iron in particular, have now led to availability of a number of specific techniques based on neutron activation analysis and mass spectrometry. With regard to measurement of stable isotopes of iron, MS provides a broader capability than NAA, but different MS techniques present their own unique advantages and limitations. Although recent developments in MS constitute major advances, the need for considerably enhanced measurement precision in isotope ratios has not yet been met. Development of a suitable MS technique capable of providing 0.1% precision for routine applications, and for all stable isotopes of iron (including $^{58}Fe$) in studies that involve a reasonable number of samples should be pursued aggressively. Currently, the only MS technique that appears to possess the ability to achieve 0.1% measurement precision consistently is TI-MS using a magnetic sector analyzer, which, unfortunately, has severe sample throughput limitations in its current design.

## 6.5 Selected Applications

With the current state of measurement methodology, i.e., routine measurements of various isotope ratios with overall precision of one percent, a number of important applications can be pursued, while a number of others must await further major improvements in measurement precision. Brief descriptions of selected applications are presented below as examples of investigations that can now be carried out to help resolve outstanding issues important in iron metabolism.

### 6.5.1 Relationship between Mucosal Absorption and Hemoglobin Incorporation of Dietary Iron

An important aspect of iron metabolism is the relationship between mucosal uptake of ingested iron and its subsequent incorporation into hemoglobin.[31,32] This relationship may have important implications in such states as prematurity, disorders of erythropoiesis, iron overload conditions, etc. For example, Ehrenkranz et al. have investigated the relationship between mucosal uptake and hemoglobin incorporation in premature infants.[9] They measured mucosal uptake of $^{58}$Fe-ferrous sulfate administered with infant formula by monitoring fecal isotope balance, and determined hemoglobin incorporation two weeks after oral dosing, using ICP-MS as measurement methodology. They observed a major difference between the amount of $^{58}$Fe that appeared to have been taken up by the gastrointestinal mucosa (41.6 ± 17.6% of dose) and what was incorporated into hemoglobin after 14 days (18.5 ± 10.0% of dose), leading to hemoglobin incorporation of only 28.7 ± 22.3% of absorbed dose. Their data are consistent with an earlier report by Gorten et al. using radio-iron.[33] This work illustrates the potential use of stable isotopes of iron in studies focusing on the dynamics of iron metabolism in prematurity and how best to provide a sufficient amount of dietary iron in order to allow for the required catch-up growth of this increasingly important group of infants. From a more practical perspective, this also highlights the potential pitfalls in equating incorporation of ingested stable isotope in hemoglobin with its dietary availability. McDonald et al. have used oral and intravenous administration of two different stable isotopes of iron in order to provide a more direct estimate of true absorption of dietary label without the need for fecal isotope balance.[20]

Currently, there is consensus that the pathogenesis of hereditary hemochromatosis (HHC) involves inappropriate mucosal uptake and/or transfer of dietary iron leading to excessive organ accumulation over many years and its toxic sequelae.[31] However, whether one or both of these plays the dominant role, as well as how to counter their effect, has not yet been conclusively elucidated.[32] Because HHC is now recognized as an important public health problem, elucidation of the specific pathogenesis and determination of whether it is possible to devise lifelong strategies to counter its effect are important.[31] Applications of stable isotopes of iron may provide a unique experimental approach to help resolve this important public health problem.

### 6.5.2 Issues of Dietary Availability of Iron

The majority of the applications of stable isotopes of iron to date have focused on various aspects of dietary availability of iron, especially in the pediatric population for whom this is a particularly important nutritional problem.[4,6–9,12–15,18,20,23,24] Investigations of iron absorption in adults have been limited, undoubtedly due in part to the current limitation of the method in measuring hemoglobin incorporation in adults.[5,10,11,16,17,21] Because of low dietary availability of iron in pediatric diets and limitations regarding multiple sampling of blood, the only suitable experimental approach for these studies is hemoglobin incorporation of the stable isotope, which appears to have been used universally. Application of this approach is straightforward, and procedures for assuring sufficient enrichment of circulating RBCs are simple. The original description of the method involves ingestion of a single isotope ($^{58}$Fe) followed by measurement of its enrichment in circulating RBCs 2 weeks later.[12,25] For the calculation of absorption from these measurements, the method relies on assessment of blood volume based on anthropometric data, measured values of hematocrit, and the assumption of a known and constant factor for incorporation of absorbed isotope into circulating RBCs.[12] This procedure does not account for two significant factors: (1) both blood volume and hemoglobin incorporation factors depend on the individual subject; and (2) iron absorption is influenced by iron stores.

Recently, Kastenmayer et al. reported use of two stable isotopes, $^{57}FeSO_4$ and $^{58}FeSO_4$, added to infant formulas on four consecutive mornings in a random manner, followed by measurement of hemoglobin incorporation 2 weeks after the last isotope feeding.[18] They calculated absorption of each label using assumed value of blood volume and measured individual values of hematocrit as detailed in the original description.[12] The (arithmetic) mean for estimated absorption (percent of dose) for the two labels in nine infants was $7.89 \pm 3.88$ (1 SD) and $7.63 \pm 4.09$ for $^{57}$Fe and $^{58}$Fe, respectively. When the mean of absorption ratios for each infant is calculated, the result ($1.047 \pm 0.234$) shows a considerable improvement in terms of variability among infants, indicating that a significant portion of the observed variability in the mean of absorption for each label is due to interindividual variability. Abrams et al.

used the double-isotope approach to develop a practical method for correcting for individual variations in iron absorption related to body stores, following the established reference dose concept used with radio-iron.[4,34] They administered 5.0 mg iron as $^{57}FeSO_4$ with cow's milk to each of ten 12 to 15 month-old children, and 2 hours later a solution of 5.0 mg iron as $^{58}FeSO_4$/ascorbic acid (Reference Dose) with grape flavoring. They estimated absorption of each label with the same procedure as described by Fomon et al.[12] The (arithmetic) means of iron absorption (percent of dose) for the $^{57}Fe$ and $^{58}Fe$ for the ten children were 5.7 ± 4.0 (1 SD) and 13.7 ± 6.4, respectively. The mean of individual ratios for absorption of $^{58}Fe$ to $^{57}Fe$ was 2.7 ± 0.9. Thus, inclusion of the reference dose reduced overall variability markedly, as is expected.[34] These two reports thus establish the utility of the double-isotope method for accurate estimation of iron absorption using the reference dose protocol to account for individual differences in iron absorption related to its status.

The report by Ehrenkranz et al. showing incomplete hemoglobin incorporation of ingested stable isotope of iron in premature infants highlights a potentially important source of error (and variability) in studies concerned with dietary availability of iron in such population groups.[9] The methodological component of this issue was addressed by McDonald et al. recently by introducing a triple-isotope method in which enriched $^{57}Fe$ and $^{54}Fe$ were used as dietary markers, and the extent of hemoglobin incorporation of absorbed label was evaluated by injecting a dose of $^{58}FeSO_4$.[20] These authors reported that in 13 premature infants, 68.3 ± 9.6% (mean ± 1 SD) of the injected label was incorporated into hemoglobin 2 weeks later. Ehrenkranz et al. reported much lower incorporation (28.7 ± 22.3) when stable isotope was ingested, using fecal isotope balance for estimating absorption of the dose.[9] Zlotkin et al. reported 18% incorporation of stable isotope in premature infants, and Gorten et al. reported 52.3 ± 28.1% incorporation using radio-iron.[33,35] These differences could be related to differences in clinical status of the infants, such as body weight, iron status, state of effectiveness of erythropoiesis, etc.[9,20,33,34] From the point of view of methodology, however, these reports indicate that injection of one of the stable isotopes can be used for direct measurement of the extent of hemoglobin incorporation, which eliminates the potentially significant source of error in the measurement of iron absorption in the patients for whom erythropoiesis is markedly incomplete.

## 6.6 Conclusion

During the past three decades, concerted efforts have been made by a number of groups to develop suitable stable isotope approaches for studies of various aspects of iron metabolism in man. While still in their early stages of application (certainly compared with radio-iron methods) these developments are very encouraging and have set the stage for future exploration of a number of important issues of iron metabolism in groups in whom the use of radioiron is not acceptable, and for future development of suitable approaches in the management of important disorders of iron metabolism such as hemochromatosis. At least three stable isotopes of iron can be used for various purposes so that the stable isotope approach allows accurate assessment of iron absorption and hemoglobin incorporation in patients in whom stable isotope backgrounds do not preclude meaningful investigations. However, the current measurement precision of the analytical methods, in comparison with the natural abundance of stable isotopes of iron, and the relatively high cost of the required dose continue to pose serious limitations for widespread applications of this approach. Major additional advances can be made only if precision of mass spectrometric methods could be further improved, and/or suitable and minimally invasive protocols could be developed for sampling the plasma pool. In order to achieve the potential for broad applications, the analytical method must provide routine isotope ratio measurements with precision of 0.1% or better and with reasonable sample throughput and overall costs — an objective for which achievement does not seem likely with the existing instrumentation and in the near future.

## References

1. Oak Ridge National Laboratory, Isotope Sales Division, Oak Ridge, TN, U.S.A., 1999.
2. Finch, C.A. et al., Ferrokinetics in man, *Medicine,* 49, 17, 1970.
3. Janghorbani, M. and Young, V. R., Use of stable isotopes to determine bioavailability of minerals in human diets using the method of fecal monitoring, *Am. J. Clin. Nutr.,* 33, 2021, 1980.
4. Abrams, S.A. et al., Absorption by 1-year-old children of an iron supplement given with cow's milk or juice, *Pediatr. Res.,* 39, 171, 1996.
5. Barrett, J.F. et al., Absorption of non-haem iron from food during normal pregnancy, *BMJ,* 309, 79, 1994.
6. Davidsson, L. et al., Iron bioavailability studied in infants: the influence of phytic acid and ascorbic acid in infant formulas based on soy isolate, *Pediatr. Res.,* 36, 816, 1994.
7. Davidsson, L. et al., Bioavailability in infants of iron from infant cereals: effect of dephytinization, *Am. J. Clin. Nutr.,* 65, 916, 1997.
8. Davidsson, L. et al., Influence of ascorbic acid on iron absorption from an iron-fortified, chocolate-flavored milk drink in Jamaican children, *Am. J. Clin. Nutr.,* 67, 873, 1998.
9. Ehrenkranz, R.A. et al., Iron absorption and incorporation into red blood cells by very low birth weight infants: studies with the stable isotope $^{58}$Fe, *J. Pediatr. Gastroenterol. Nutr.,* 15, 270, 1992.
10. Fairweather-Tait S.J., Minski, M.J., and Richardson, D.P., Iron absorption from a malted cocoa drink fortified with ferric orthophosphate using the stable isotope $^{58}$Fe as an extrinsic label, *Br. J. Nutr.,* 50, 51, 1983.

11. Fairweather-Tait, S.J. and Minski, M.J., Studies on iron availability in man, using stable isotope techniques, *Br. J. Nutr.*, 55, 279, 1986.
12. Fomon, S.J. et al., Erythrocyte incorporation of ingested 58-iron by infants, *Pediatr. Res.*, 24, 20, 1988.
13. Fomon, S.J. et al., Iron absorption from infant foods, *Pediatr. Res.*, 26, 250, 1989.
14. Fomon, S.J., Ziegler, E.E., and Nelson, S.E, Erythrocyte incorporation of ingested $^{58}$Fe by 56-day-old breast-fed and formula-fed infants, *Pediatr. Res.*, 33, 573, 1993.
15. Fomon, S.J. et al., Erythrocyte incorporation of iron is similar in infants fed formulas fortified with 12 mg/L or 8 mg/L of iron, *J. Nutr.*, 127, 83, 1997.
16. Hashimoto, F. et al., Determination of absorption and endogenous excretion of iron in man by monitoring fecal excretion of a stable isotope ($^{58}$Fe), *J. Nutr. Sci. Vitaminol.* (Tokyo), 38, 435, 1992.
17. Janghorbani, M., Ting, B.T., and Young, V.R., Absorption of iron in young men studied by monitoring excretion of a stable iron isotope ($^{58}$Fe) in feces, *J. Nutr.*, 110, 2190, 1980.
18. Kastenmayer, P. et al., A double isotope technique for measuring iron absorption in infants, *Br. J. Nutr.*, 71, 411, 1994.
19. Lehman, W.D., Fischer, R., and Heinrich, H.C., Iron absorption in man calculated from erythrocyte incorporation of the stable isotope iron-54 determined by fast atom bombardment mass spectrometry, *Anal. Biochem.*, 172, 151, 1988.
20. McDonald, M.C., Abrams, S.A., and Schanler, R.J., Iron absorption and red blood cell incorporation in premature infants fed an iron-fortified infant formula, *Pediatr. Res.*, 44, 507, 1998.
21. Minihane, A.M. and Fairweather-Tait, S.J., Effect of calcium supplementation on daily nonheme-iron absorption and long-term iron status, *Am. J. Clin. Nutr.*, 68, 96, 1998.
22. Van den Heuvel, E.G. et al., Nondigestible oligosaccharides do not interfere with calcium and nonheme-iron absorption in young, healthy men, *Am. J. Clin. Nutr.*, 67, 445, 1998.
23. Woodhead, J.C. et al., Use of stable isotope, $^{58}$Fe, for determining availability of nonheme iron in meals, *Pediatr. Res.*, 23, 495, 1988.
24. Woodhead, J.C. et al., Gender-related differences in iron absorption by preadolescent children, *Ped. Res.*, 29, 435, 1991.
25. Janghorbani, M., Ting, B.T., and Fomon, S.J., Erythrocyte incorporation of ingested stable isotope of iron ($^{58}$Fe), *Am. J. Hematol.*, 21, 277, 1986.
26. Miller, D.D. and van Campen, D., A method for the detection and assay of iron stable isotope tracers in blood serum, *Am. J. Clin. Nutr.*, 32, 2354, 1979.
27. Ting, B.T. et al., Radiochemical neutron activation analysis of stable isotopes in relation to human mineral nutrition, *J. Radioanal. Chem.*, 70, 133, 1981.
28. Ting, B.T. and Janghorbani, M., Inductively coupled plasma mass spectrometry applied to isotopic analysis of iron in human fecal matter, *Anal. Chem.*, 58, 1334, 1986.
29. Kim, H.W., Yu, Y.J., and Greenburg, A.G., Iron-58 and neutron activation analysis: a non-radioactive method for tracing hemoglobin iron, *Artif. Cells Blood Substit. Immobil. Biotechnol.*, 22, 619, 1994.
30. Eagles, J., Fairweather-Tait, S.J., and Self, R., Stable isotope ratio mass spectrometry for iron bioavailability studies, *Anal. Chem.*, 57, 469, 1985.
31. Bacon, B.R. and Brown, K.E., Iron metabolism and disorders of iron overload, in *Liver and Biliary Diseases*, Second edition, Williams & Wilkins, Baltimore, 1996, 349.

32. McLaren, G.D. et al., Regulation of intestinal iron absorption and mucosal iron kinetics in hereditary hemochromatosis, *J. Lab. Clin. Med.*, 117, 390, 1991.
33. Gorten, M.K., Hepner, R., and Workman, J.B., Iron metabolism in premature infants. I. Absorption and utilization of iron as measured by isotope studies, *J. Ped.*, 63, 1063, 1963.
34. Cook, J.D. et al., Food iron absorption measured by an extrinsic tag. *J. Clin. Invest.*, 51, 805, 1972.
35. Zlotkin, S.H. et al., Determination of iron absorption using erythrocyte iron incorporation of two stable isotopes of iron ($^{57}Fe$ and $^{58}Fe$) in very low birthweight premature infants, *J. Ped. Gastroenterol. Nutr.*, 21, 190, 1995.

# 7

# *Use of Isotopes in the Assessment of Zinc Status*

**Malcolm J. Jackson and Nicola M. Lowe**

**CONTENTS**

7.1 Introduction ... 109
7.2 Isotopic Techniques ... 111
    7.2.1 Short-term Two-compartment Model ... 112
    7.2.2 Simplified Techniques to Measure the Exchangeable Zinc Pool ... 113
7.3 Conclusion ... 113
References ... 114

## 7.1 Introduction

Zinc is an essential trace element that plays a crucial role in many aspects of biochemistry, acting variously as a co-factor for enzymes, an essential component of zinc "finger" transcription factors, and structural component of many molecules.[1] Zinc deficiency has been recognized in animals since the 1950s and evidence for its occurrence in man was first provided by Prasad and coworkers, although the validity of these early reports has subsequently become the subject of considerable debate and contention.[2–4] Unequivocal, acute deficiency was finally demonstrated by Moynahan, while examining children with acrodermatitis enteropathica, a severe, inherited dermatological disorder.[5] Subsequently, there have been many reports of zinc deficiency in populations and individuals in various parts of the world, although, in the absence of clear responses to zinc supplementation, many of these remain contentious.[6] Much of the confusion in this area stems from a lack of reliable biochemical and clinical indicators of mild zinc depletion.[7] In the chronic mild form, zinc deficiency can lead to many non-specific features, such as

0-8493-8730-2/00/$0.00+$.50

**TABLE 7.1**

Techniques Used to Assess Body Zinc Status

| |
|---|
| Measurement of the zinc content of accessible body fluids |
| Serum/plasma |
| Urine |
| Measurement of the zinc content of accessible tissues |
| Measurement of the activity of zinc-dependent enzymes |
| Measurement of the uptake of radioactive zinc by isolated cells |
| Measurement of the expression of specific proteins dependent upon zinc for normal function (e.g., metallothionein, thymulin) |
| Assessment of specific responses to zinc supplementation |
| Measurement of zinc flux rates with isotopes |
| Measurement of body pool sizes with isotopes |

Derived from Reference 7.

growth retardation or an immune deficit; it can also exacerbate existing diarrheal disease, etc. — features which are not diagnostic of the lack of zinc.

Many different approaches have been followed to try to develop reliable indicators of zinc status. The general groups of techniques are shown in Table 7.1. There are a number of specific problems with most of the approaches shown. Numerous studies now clearly show that a fall in dietary zinc intake leads to a rapid fall in the concentration of zinc in extracellular fluids (e.g., plasma or serum). Thus, zinc deficiency is usually associated with a fall in plasma or serum zinc content. Plasma zinc levels also fall, however, in response to non-specific stress conditions unassociated with zinc deficiency, most notably in association with bacterial infections and in women who are pregnant or using oral contraceptive agents.[8] A finding of low plasma or serum zinc is therefore not immediately diagnostic of body zinc depletion. In contrast, urine zinc levels are unresponsive to dietary zinc levels except in situations where dietary intakes decrease to very low levels. Relatively few studies have examined the zinc content of accessible tissues to evaluate whole-body zinc status. Zinc is primarily an intracellular ion with greater than 85% of body zinc found within muscle and bone (Table 7.2). Readily accessible tissues include erythrocytes, white blood cells and hair, in addition to blood serum, but none have been proven to provide reliable routine methods for the analysis of tissue zinc status.[9,10]

Despite numerous attempts to find a single zinc-responsive enzyme from among the 300-plus that are dependent upon zinc, none have been shown to be reliable indicators.[9] Some investigators have claimed that measurements of alkaline phosphatase activities may be a useful index of zinc status, but recent data indicate that changes in activity are relatively insensitive to changes in zinc status (Lowe et al., unpublished observations). In addition, the possibility of non-specific changes in plasma or serum activities, due to changes in liver or bone turnover or damage, argue against this proving a robust routine marker of zinc status.

There is considerable interest in the possibility that measurement of metallothionein (MT) protein or mRNA levels in blood cells might provide a reliable

**TABLE 7.2**

Distribution of Zinc in the Human Body

| Tissue | Approximate Zinc Concentration (μg/g. wet wt.) | Total Zinc Content (g) | Proportion of Total Body Zinc (%) |
|---|---|---|---|
| Skeletal muscle | 51 | 1.53 | 57(approx.) |
| Bone | 100 | 0.77 | 29 |
| Skin | 32 | 0.16 | 6 |
| Liver | 58 | 0.13 | 5 |
| Brain | 11 | 0.04 | 1.5 |
| Kidneys | 55 | 0.02 | 0.7 |
| Heart | 23 | 0.01 | 0.4 |
| Hair | 150 | <.01 | 0.1 (approx.) |
| Blood plasma | 1 | <.01 | 0.1 (approx.) |

From Reference 10.

adjunct assay for measurement of zinc status.[9] Again, despite initial enthusiasm for the approach, the specificity of this assay has been questioned as investigators have identified multiple factors influencing MT expression in blood cells.[11]

Despite numerous investigations, reliable routine measures of zinc status remain elusive. Data from a number of sources have indicated that functional zinc deficiency becomes apparent with loss of only a small proportion of total body zinc.[10,12] This relatively small mobilizable, or functional "pool" of zinc, appears to contribute less than ten percent of total body zinc. Clearly, any reliable indicator of zinc status must be responsive to changes in the magnitude of this pool.

Isotopic techniques provide a route to access these pools, potentially providing a means of measuring their size and turnover. Additionally, isotopes can be used to provide information on parameters of zinc metabolism that are relevant to the assessment of zinc status, including fractional rates of gastrointestinal zinc absorption and the rate of gastrointestinal excretion/secretion of zinc (see Chapter 5).

## 7.2 Isotopic Techniques

Both radio- and stable-isotopes have been used for studies of zinc status. Details of the isotopes available, radioactive half-life and natural enrichments are provided in other chapters.

Few investigators now use radioactive zinc isotopes, but previous studies with $^{65}$Zn have provided considerable information on body zinc turnover. The long half-life and γ-emission of this isotope have facilitated extended turnover studies with surface counting of the isotopic activity in specific organs.[13] This detailed information was used by Wastney et al. to develop an extremely comprehensive mathematical model of whole-body zinc metabolism in man, which remains the "gold standard" for studies in this area.[23] The model provides information and turnover of zinc in multiple-body pools for control subjects against which data from test groups can be compared. Coincident with establishment of this model has been the development of "user-friendly" computer-based modelling software (e.g., SAAM/CONSAAM) which has facilitated use of complex models, such as the model of Wastney et al. and development of alternatives based upon and fully compatible with it .[14,22,23]

Despite the large amount of data that can readily be obtained with radioactive isotopes, most investigators have concluded that such studies cannot be justified due to the radiation doses involved. Most recent data, therefore, have been obtained with stable isotopes of zinc, usually $^{67}Zn$ or $^{70}Zn$. Studies specifically designed to investigate exchangeable zinc pool sizes were undertaken by Jackson et al.[15,16] These assumed that a simple one-compartment dilution was applicable to intravenously administered $^{67}Zn$, a situation that is now known to be incorrect. Nevertheless, these studies demonstrated the potential applicability of the approach.

A number of approaches have subsequently been followed that used stable isotopes and computer modelling of data to derive pool size and turnover rates. Two approaches will be described here, together with some discussion of their advantages, disadvantages, and applicability.

### 7.2.1 Short-term Two-compartment Model

Lowe et al. examined the plasma decay of $^{70}Zn$ following intravenous injection of 0.5 mg of 96.5% enriched $^{70}Zn$ as $ZnCl_2$.[17] They found that over a period of 120 minutes the decay closely followed bi-exponential kinetics and the data could be used to derive a two-compartment model of short-term plasma zinc kinetics (Figure 7.1). Comparison of these data with previous studies on animals indicated that the initial zinc pool was primarily contributed by the blood plasma, while the second pool was primarily a rapidly exchanging portion of the liver zinc.[18,19] Lowe et al. used this approach to study the zinc status of a group of patients with alcoholic cirrhosis, concluding that they had significant abnormalities in zinc metabolism.[17]

This limited, short-term analysis of isotope turnover makes many assumptions concerning the rapid distribution of isotopic zinc in the body and may not be applicable in all situations. It is reasonably rapid and simple, requiring a single isotope administration and relatively few blood samples. It can be used to help differential diagnosis of zinc deficiency where serum zinc levels have initially been shown to be low, and in which a stress-related fall is potentially involved.[18,20] The approach has been extended with monitoring of serum isotope enrichment for longer periods and addition of further pools to the model, but the reliability of these developments has not been evaluated.[21,22]

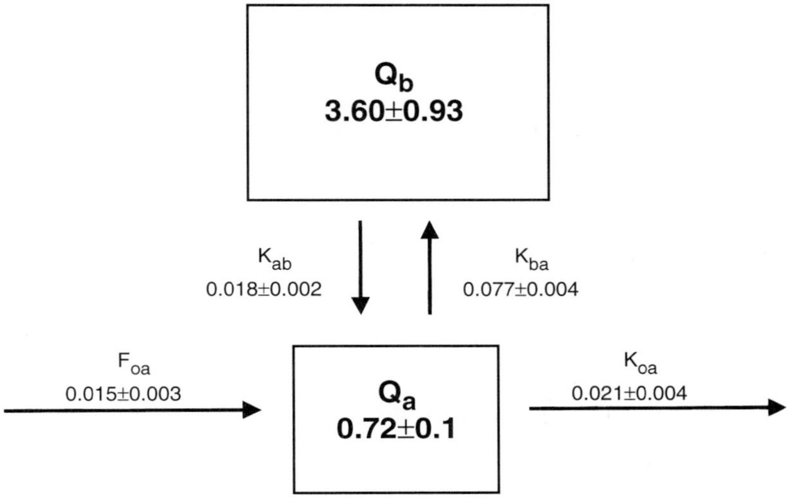


**FIGURE 7.1**
Two-compartment model of plasma zinc kinetics in man. Schematic representation of the two-compartment model of plasma zinc kinetics showing the initial (a) and second (b) pools (μmol/kg) with which intravenously injected $^{70}Zn$ equilibrated in control subjects. The fractional rates of transfer-per-minute are also shown. (From Lowe et al., 1993.)

### 7.2.2 Simplified Techniques to Measure the Exchangeable Zinc Pool

In order to develop simpler techniques, capable of use in greater numbers of subject groups, Hambidge and co-workers have derived information from the comprehensive model published by Wastney et al. to attempt to determine the magnitude of the body "exchangeable zinc pool" or EZP.[22,23] This group of compartments is defined as those that exchange with plasma zinc within 2 days and is calculated following either oral or intravenous administration and monitoring of the plasma or urine isotope enrichment between 3 and 9 days after isotope administration.[22] Comparison with more complex models indicates that this simplified technique overestimates the amount present by approximately 20%.[22] This approach has now been used to determine zinc status in several patient and subject groups, including infants.[1,24] The data appear robust and provide a useful adjunct to conventional indicators of zinc status.

## 7.3 Conclusion

It is now clear that, although the use of isotopes can aid in the assessment of zinc status in man, the techniques are not practical for widespread use in population studies because of the cost of isotopes and their analysis, the necessity for specialized equipment, and the labor intensity of the protocols. Simplified protocols, such as that proposed by Miller et al. may aid in this, but further validation is necessary.[22] Fundamentally, the need for subjects to be in a "steady state" throughout the duration of the study may also limit the applicability of modelling-based approaches.

## References


1. Krebs, N.F. et al., The use of stable isotope techniques to assess zinc metabolism, *Nutr. Biochem.*, 6, 292, 1996.
2. Prasad, A.S., Halsted, J.A., and Nadami, M., Syndrome of iron deficiency anaemia, hepatosplenomegaly, hypogonadism, dwarfism and geophagia, *Am. J. Med.*, 31, 532, 1961.
3. Coble, Y.D., Schulert, A.R., and Farid, Z. Growth and sexual development of male subjects in an Egyptian oasis, *Am. J. Clin. Nutr.*, 18, 421, 1969.
4. Coble, Y.D., Van Reem, R., and Schulert, A.R., Zinc levels and blood enzyme activities in Egyptian male subjects with retarded growth and sexual development, *Am. J. Clin. Nutr.*, 18, 421, 1966b.
5. Moynahan, E.J., Acrodermatitis enteropathica: a lethal inherited human zinc deficiency disorder, *Lancet ii*, 3999, 1974.
6. Mills, C., *Zinc in Human Biology*, Springer-Verlag, London, 1988.
7. Jackson, M.J. and Lowe, N.M., *Trace Element Metabolism in Man and Animals — 7*. IMI Zagreb, 1991, 4.3.
8. Halbrook, R. and Hedelin, H., Zinc metabolism and surgical trauma. *Br. J. Surg.*, 64, 271, 1977.
9. Golden, M.H.N., *Zinc in Human Biology*, Springer-Verlag, London, 1988, 323.
10. Jackson, M.J., *Zinc in Human Biology*, Springer-Verlag, London, 1988, 1.
11. Morrison, J.M., Wood, A.M., and Bremner, I., Effects of protein deficiency on blood cell and tissue metallothionein-1 concentrations in rats, *Proc. Nutr. Soc.*, 49, 68A, 1990.
12. Jackson, M.J., Jones, D.A., and Edwards R.H.T., Tissue zinc levels as an index of body zinc status, *Clin. Physiol.*, 2, 333, 1982.
13. Foster, D.M. et al., Zinc metabolism in humans: a kinetic model. *Am. J. Physiol.*, 237, R340, 1979.
14. Lowe, N.M. et al., Estimation of zinc absorption in humans using four stable isotope tracer methods and compartmental analysis, *Am. J. Clin. Nutri.*, 71, 523, 2000.
15. Jackson, M.J. et al., Zinc homeostasis in man: Studies using a new stable isotope dilution technique, *Br. J. Nutr.*, 51, 199, 1984.
16. Jackson, M.J. et al., Stable isotope metabolic studies of zinc nutrition in slum-dwelling lactating women in the Amazon valley, *Br. J. Nutr.*, 59, 193, 1988.
17. Lowe, N.M. et al., Studies of human zinc kinetics using the stable isotope $^{70}$zinc, *Clin. Sci.*, 84, 113, 1993.
18. Lowe, N.M., Bremner, I., and Jackson, M.J., Plasma 65-zinc kinetics in the rat, *Br. J. Nutr.*, 65, 445, 1991.

19. Chesters, J.K. and Will, M., Measurement of the flux through plasma in normal and endotoxin-stressed pigs and the effects of zinc supplementation during stress, *Br. J. Nutr.*, 46, 119, 1981.
20. Lowe, N.M. et al., A stable isotope study of zinc kinetics in Irish Setters with gluten sensitive enteropathy, *Br. J. Nutr.*, 74, 69, 1995.
21. Fairweather-Tait, S. et al., The measurement of exchangeable pools of zinc using the stable isotope $^{70}Zn$, *Br. J. Nutr.*, 70, 221, 1993.
22. Miller, L.V. et al., Size of the zinc pools that exchange rapidly with plasma zinc in humans: alternative techniques for measuring and relation to dietary zinc intake, *J. Nutr.*, 124, 268, 1994.
23. Wastney, M. et al., Kinetic analysis of zinc metabolism and its regulation in normal humans, *Am. J. Physiol.*, 251 R398, 1986.
24. Sian, L. et al., Zinc absorption and intestinal losses of endogenous zinc in young Chinese women with marginal zinc intakes, *Am. J. Clin. Nutr.*, 63, 348, 1996.

# 8

# *Copper Status and Metabolism Studied with Isotopic Tracers*

**Judith R. Turnlund**

**CONTENTS**

8.1 Introduction ........ 117
8.2 Background ........ 118
8.3 Copper Status ........ 118
8.4 Isotopic Tracers ........ 119
 8.4.1 Radioactive Tracers ........ 119
 8.4.2 Stable-isotope Tracers ........ 120
  8.4.2.1 Methods of Stable-isotope Analysis ........ 120
   8.4.2.1.1 Neuron Activation Analysis ........ 120
   8.4.2.1.2 Electron Impact Mass Spectrometry and Gas Chromatography Mass Spectrometry .. 120
   8.4.2.1.3 Thermal Ionization Mass Spectrometry .... 121
   8.4.2.1.4 Inductively Coupled Plasma Mass Spectrometry ........ 121
  8.4.2.2 Multiple Stable-isotope Labelling ........ 121
  8.4.2.3 Studies Using Isotopic Tracers of Copper ........ 122
8.5 Conclusion ........ 123
References ........ 123

## 8.1 Introduction

The use of isotopic tracers for research on copper in humans has resulted in new, definitive information that aids in developing an understanding of the metabolism of copper. In recent years, most studies of copper metabolism in humans have been conducted with stable-isotope tracers, but radioisotopes were used in early tracer studies and in select recent studies. Isotopic tracers

0-8493-8730-2/00/$0.00+$.50

can be used in conjunction with traditional indices of copper status to gain a better understanding of a variety of areas related to copper nutriture.

## 8.2 Background

Copper is an essential element for humans as well as animals. Studies of copper metabolism in humans began in the 1930s. Even though there was no known copper deficiency in humans, an estimate of the copper requirement for women was made in 1944.[1–3]

Copper deficiency in humans was discovered in the 1960s.[4,5] Deficiency has been observed in humans under a number of special circumstances. Conditions that have led to deficiency symptoms include patients on long-term parenteral nutrition without copper added to the parenteral solution, infants recovering from protein-energy malnutrition, and premature infants fed cow's milk.[6] Manifestations of frank copper deficiency include anemia not responsive to iron supplementation, leukopenia, and neutropenia. Osteoporosis is observed when bones are still growing.[7]

Genetic defects in copper metabolism have been identified in humans. Menkes' disease results in copper deficiency due to a defect in copper transport. Wilson's disease results in excessive copper storage, as does childhood cirrhosis.[4]

## 8.3 Copper Status

Inadequate copper status is easily established in cases of frank copper deficiency; serum copper and ceruloplasmin levels are very low.[8] However, it is not clear whether these indices of copper status are as useful when copper intake is marginal. In addition, these parameters are increased in a number of conditions, which could mask copper deficiency. Serum copper and ceruloplasmin rise markedly during pregnancy, following surgery, and with inflammatory conditions, infections, diabetes, coronary and cardiovascular diseases, uremia and malignant diseases.[4,9]

A number of other biochemical indicators of copper status have been considered. Erythrocyte superoxide dismutase activity is considered by some to be preferable to serum copper and ceruloplasmin.[10] Copper levels in hair, nails, or saliva have been suggested as indicators, but do not appear to reflect copper status. Urinary copper declines when dietary copper is very low, but does not otherwise relate to dietary copper intake.[11] Recent studies suggest that platelet copper and platelet cytochrome c oxidase activity may respond more quickly to copper depletion than the indicators above.[12] Leukocyte copper also declines with copper depletion, but little data are available on this parameter.[13] Studies in laboratory animals and in patients with Menkes' disease, a condition resulting in severe copper deficiency due to a defect in copper transport, suggest peptidyl glycine $\gamma$-aminating monooxygenase may be a diagnostic tool for copper deficiency.[14] Copper balance has been used for establishing dietary recommendations in the past, but there are a number of problems with the balance approach.[4,15] Balance can be achieved over a wide range of intakes; a sufficient adaptation period is required for balance data to be meaningful. Miscellaneous losses of copper contribute to balance data and a number of sources of error affect the reliability of the data.

The limitations of traditional and potential biochemical indices of copper status have led to exploring the potential of isotopic tracers to aid in evaluating copper status. Isotopic tracers offer a number of unique features that make them important to the study of copper metabolism. They are particularly valuable in the study of the metabolic fate of copper, as well as studies of copper absorption, utilization, excretion, and turnover.

## 8.4 Isotopic Tracers

Studies of copper metabolism using radioactive tracers began in laboratory animals in the 1940s,[16] and were followed by studies of copper metabolism in humans, using radioisotopes. A discussion of these studies is covered in the section on radioisotopic tracers below. When the limitations of radioisotopes became apparent, interest in using stable isotopes of copper was stimulated.[17] The limitations of stable isotopes of copper, their advantages, and use in metabolic research are discussed below.

### 8.4.1 Radioactive Tracers

There are seven radioisotopes of copper, but all have relatively short half-lives.[18] The two isotopes with the longest half-lives are $^{64}Cu$, a beta emitter with a half-life of 12.8 hours, and $^{67}Cu$, a beta and gamma emitter with a half-life of 58.5 hours. Both are used in metabolic research. However, their use is subject to limitations. Because of the short half-lives, long-term studies would require relatively high doses of the isotopes and result in unacceptable levels of exposure to radiation. In some situations, especially those infants and pregnant women, exposure to any radiation, no matter how low, is not acceptable.

Radioisotopes are used primarily in studies in laboratory animals and have provided valuable information on copper metabolism. These include research on absorption, excretion and distribution, and kinetics.[19–23]

Studies of copper metabolism using radioactive tracers in humans began in the 1950s. Two examples of these studies are research on copper metabolism in hepatolenticular degeneration, and transfer of copper between erythrocytes and plasma.[24,25] A number of research studies on Wilson's disease have made use of radioisotopes of copper.[26–31] Using radioisotopes, copper metabolism has also been investigated in patients with biliary cirrhosis.[32,33]

Recent advances in technology for whole-body counting of gamma emitters have allowed detection of extremely low levels of $^{67}Cu$, and made it possible to study copper absorption in healthy subjects with very low exposure to radiation.[34,35] This approach was used to study absorption of copper and two other gamma emitters, zinc and iron, simultaneously.

### 8.4.2 Stable-isotope Tracers

Copper has two stable isotopes and both are relatively abundant. $^{65}Cu$ has a natural abundance of 30.8% and $^{63}Cu$ has an abundance of 69.2%.[36]

Ideally, stable isotopes used as tracers have low natural abundance, but analytical methods are sufficiently sensitive that stable-copper isotopes are very effective tracers. However, they are used in amounts higher than traditional "tracer" doses. When an element has only one stable isotope, it cannot be used as a tracer. Since copper has two isotopes, it can be used. The applications, however, are more limited than minerals such as molybdenum, with seven stable isotopes, or even magnesium, with three stable isotopes. Only one stable isotope of copper can be enriched at one time, since it must be compared to an isotope with a known abundance. Multiple isotopes can be enriched simultaneously when a mineral has multiple stable isotopes, as has been demonstrated with molybdenum and a number of other minerals.[37]

#### *8.4.2.1 Methods of Stable-isotope Analysis*

A number of analytical methods have been used to measure isotopic ratios of copper in samples from human nutrition studies. These include neutron activation analysis (NAA), electron ionization mass spectrometry (EIMS), gas chromatography-mass spectrometry (GCMS), thermal ionization mass spectrometry (TIMS), and inductively coupled plasma mass spectrometry (ICP-MS). Recent work has been primarily with TIMS and ICP-MS.

##### *8.4.2.1.1 Neutron Activation Analysis*

NAA was the first method applied to the analysis of stable isotopes of copper in nutrition research.[17] It was used in early studies of copper nutriture in humans, but the approach has not been widely used, due to the limitations of the method.[38,39] It has relatively poor precision for copper compared to other methods; in addition, it requires the availability of a nuclear reactor.

##### *8.4.2.1.2 Electron Impact Mass Spectrometry and Gas Chromatography Mass Spectrometry*

EIMS and GCMS are methods used often in chemistry laboratories and for biochemical research to measure organic compounds. Chelation of the element to be measured is required to convert it to a volatile organic compound. The method has been explored as a way to measure stable isotopes of copper. The approach was used in early stable-isotopes studies.[40–43] Both EIMS and GCMS are susceptible to interference, and inadequate precision has limited their use for measuring of stable isotopes of copper.

##### *8.4.2.1.3 Thermal Ionization Mass Spectrometry*

TIMS is the method capable of the highest precision and accuracy for the measurement of isotopic ratios of copper.[44] It was used for the determination of the atomic weight of copper, as well as of most other minerals.[36] The method was used in geochemistry and nuclear chemistry long before it was applied to studies of copper metabolism. Early instruments were built in the laboratories of mass spectrometrists and thus not widely available, but, in the 1980s, commercial instruments became available, making the instrumentation available in other laboratories. Most TIMS instruments have a magnetic sector analyzer, but a few use a quadrupole. The quadrupole instruments are not capable of the high precision and accuracy of the magnetic sector instruments; therefore, use for copper has been limited.

While capable of the highest precision, there are a number of drawbacks to magnetic sector TIMS. Copper must be separated from the sample matrix. A high degree of purity is required for analysis. After eliminating the organic matrix, copper must be separated from other minerals in the sample, usually by anion exchange chromatography.[45] The separation procedures are slow and require considerable care to minimize contamination. Analysis is relatively slow and only about 12 samples can be analyzed a day. Most of the research to date on copper metabolism in humans has employed TIMS for analysis.

##### *8.4.2.1.4 Inductively Coupled Plasma Mass Spectrometry*

ICP-MS is the newest approach to measuring isotopic ratios of copper for metabolic studies. A detailed description of the method has been reported.[40] While few reports of studies using ICP-MS for copper stable-isotope studies have been published yet, the advantages of the method suggest it will be the method of choice in the future. It offers better precision than all methods except magnetic sector TIMS. The advantages over TIMS are that less laborious sample preparation is required, sample throughput is faster, and samples containing less copper can be analyzed.

### ***8.4.2.2 Multiple Stable-isotope Labelling***

A major advantage of stable isotopes is that isotopes of a number of elements can be administered simultaneously. As a result, interactions between several minerals can be investigated simultaneously. In our laboratory, we have administered isotopes of as many as five minerals simultaneously.[46] This has included various combinations of isotopes of the minerals copper, molybdenum, zinc, iron, calcium, and magnesium.[45–47]

The total amount of a mineral in samples from isotopic studies is often measured by a method such as atomic absorption spectrophotometry. However, isotope dilution can be used for determining the quantity of the element of interest and the enriched isotope appearing in a sample. Since copper has only two stable isotopes, isotope dilution is done by analyzing the isotopic ratios of a sample, spiking a duplicate sample with a stable isotope, and determining the isotopic ratios of that sample.[48] This requires analyzing two samples, rather than one, by mass spectrometry, but offers the advantage of determining the quantity of both the isotope and the mineral by the same high precision method. This reduces the bias introduced by using one method for quantifying the mineral and another to quantify the isotope. The derivation and calculations for both methods have been reported.[48,49] The isotope dilution calculations are shown below.

The total mineral content ($M^m$) and the enriched isotope content ($M^s$) of samples were calculated using:

- The isotopic ratios of the enriched sample collected before and after addition of the isotopic diluent.
- Unenriched samples and the isotopically enriched solution.
- The weights of the sample and the isotopic diluent added to the sample.
- The concentration of the isotopic diluent.
- The total dry weight of the sample in the equations below.

$$M^s = \frac{M^d \cdot (R^m{}_{jk} - R^n{}_{jk})(R^s{}_{jk} - R^t{}_{jk})}{f \cdot (R^t{}_{jk} - R^m{}_{jk})(R^s{}_{jk} - R^n{}_{jk})} \tag{8.1}$$

$$M^n = \frac{M^d \cdot A^s{}_k \cdot W^n (R^m{}_{jk} - R^n{}_{jk})(R^s{}_{jk} - R^t{}_{jk})}{f \cdot A^n{}_k \cdot W^s (R^t{}_{jk} - R^m{}_{jk})(R^s{}_{jk} - R^n{}_{jk})} \tag{8.2}$$

$$M^m = M^n + M^s \tag{8.3}$$

where: M = mass of copper, A = isotopic abundance, W = atomic mass, R = isotopic ratio, f = fraction of sample weighed for analysis, j = isotope enriched ($^{65}$Cu), k = reference isotope ($^{63}$Cu), n = natural copper (unenriched), s = isotopically enriched copper fed to subjects, d = isotopically enriched copper used for isotope dilution, m = mixture of n and s in total sample, t = mixture of m and d in sample analyzed.

### *8.4.2.3 Studies Using Isotopic Tracers of Copper*

To date, stable isotopes of copper have been used primarily to determine copper absorption and bioavailability. Copper absorption has been measured in men over a broad range of dietary copper intakes, ranging from 0.4 to 8 mg/day. These studies demonstrated that the efficiency of absorption declines as intake increases, protecting from copper deficiency and toxicity.[50–52] Comparisons have been made of copper absorption between young and elderly men, and between pregnant and non-pregnant women.[53,54] Copper absorption has been assessed in very-low-birth-weight infants.[55] The effects of dietary components on copper absorption have been compared in vegetarian and non-vegetarian diets, in diets with and without high amounts of phytate, and with a high fiber diet.[54,56,57] Interactions with other nutrients have been compared in copper absorption studies during vitamin B-6 depletion, with two levels of dietary zinc, and with a low-zinc diet.[48,58,59] The effects of age and sex on copper absorption and biological half-life have been studied with a radioisotope of copper.[60] Plants and meats have been labelled with isotopes of copper and then fed to humans to study copper absorption from specific foods, including wheat, goose meat, and peanut butter.[61–63]

Kinetic studies of copper metabolism hold promise of being an additional tool for assessing copper status. Estimates of the body pool of copper could provide information not available from biochemical tests. Kinetic models of copper metabolism have been developed in laboratory animals and in sheep, using radioisotopes.[22,23,64] Kinetic studies in dairy cows have employed stable isotopes.[65,66] Kinetic studies of copper metabolism have also begun in humans.[67,68] Continuous dosing of a stable isotope was used in rats to measure long-term copper turnover in organs.[69]

## 8.5 Conclusion

The use of isotopic tracers for studying copper metabolism has increased markedly in the past decade. A considerable amount of research has been done on copper absorption and bioavailability. Some areas, such as kinetics and turnover, have just begun to be explored with stable isotopes. While it is expected that scientists will continue to use traditional methods of assessing status, adding the new tools provided by isotopic tracers will provide valuable new information to aid in developing an understanding of the regulation of copper metabolism and maintenance of status.

## References

1. Chou, T. and Adolph, W.H., Copper metabolism in man, *Biochem. J.*, 29, 476–479, 1935.
2. Tompsett, S.L., The excretion of copper in urine and faeces and its relation to the copper content of the diet, *Biochem. J.*, 28, 2088–2091, 1934.

3. Leverton, R.M. and Binkley, E.S., The copper metabolism and requirement of young women, *J. Nutr.*, 27, 43–52, 1944.
4. Mason, K.E., A conspectus of research on copper metabolism and requirements of man, *J. Nutr.*, 109, 1979–2066, 1979.
5. Cartwright, G.E. and Wintrobe, M.M., The question of copper deficiency in man, *Am. J. Clin. Nutr.*, 14, 94–110, 1964.
6. Williams, D.M., Copper deficiency in humans, *Sem. Hemat.*, 20, 118–127, 1983.
7. Turnlund, J.R., Copper, in M.E. Shils, J.A. Olson, M. Shike, and A.C. Ross (Eds.), *Modern Nutrition in Health and Disease*, 9th Ed., pp. 241–252 Williams & Wilkins, Baltimore, 1999.
8. Fujita, M., Itakura, T., Takagi, Y., and Okada, A., Copper deficiency during total parenteral nutrition: Clinical analysis of three cases, *JPEN*, 13, 421–425, 1989.
9. Davis, G.K. and Mertz, W., Copper, *in Trace Elements in Human and Animal Nutrition*, Mertz, W., Ed., Volume 1, 5th ed., pp. 301–364 Academic Press, San Diego, 1987.
10. Uauy, R., Castillo-Duran, C., Fisberg, M., Fernandez, N., and Valenzuela, A., Red cell superoxide dismutase activity as an index of human copper nutrition, *J. Nutr.*, 115, 1650–1655, 1985.
11. Turnlund, J.R., Scott, K.C., Peiffer, G.L., Jang, A.M., Keen, C.L., and Sakanashi, T.M., Copper status of young men consuming a low copper diet, *Am. J. Clin. Nutr.*, 65, 72–78, 1997.
12. Milne, D.B. and Nielsen, F.H., Effects of a diet low in copper on copper-status indicators in postmenopausal women, *Am. J. Clin. Nutr.*, 63, 358–364, 1996.
13. Turnlund, J.R., Sakanashi, T.M., and Keen, C.L., Polymorphonuclear leukocyte copper declines with low dietary copper, in P.W.F. Fischer, M.R. L'Abbe, K.A. Cockell, and R.S. Gibson, Eds., *Trace Elements in Man and Animals — 9: Proceedings of the Ninth International Symposium on Trace Elements in Man and Animals*, pp. 113–114. NRC Research Press, Ottawa, Canada, 1997.
14. Prohaska, J.R., Tamura, T., Percy, A.K., and Turnlund, J.R., *In vitro* copper stimulation of plasma peptidylglycine a-amidating monooxygenase in Menkes' disease variant with occipital horns, *Pediatr. Res.*, 42, 862–865, 1997.
15. Mertz, W., Use and misuse of balance studies, *J. Nutr.*, 117, 1811–1813, 1987.
16. Schultze, M.O. and Simmons, S.J., Use of radioactive copper in studies on nutritional anemia of rats, *J. Biol. Chem.*, 142, 97–106, 1942.
17. Lowman, J.T. and Krivit, W., New *in vivo* tracer method with the use of nonradioactive isotopes and activation analysis, *J. Lab. Clin. Med.*, 61, 1042–1048, 1963.
18. Dyer, F.F. and Leddicotte, G.W., *The Radiochemistry of Copper*, Washington, D.C., National Research Council, National Academy of Sciences, 1961.
19. Owen, C.A., Jr., Distribution of copper in the rat, *Am. J. Physiol*, 207, 446–448, 1964.
20. Owen, C.A., Jr., Absorption and excretion of $Cu^{64}$-labeled copper by the rat, *Am. J. Physiol.*, 207, 1203–1206, 1964.
21. Owen, C.A., Jr., Metabolism of radiocopper ($Cu^{64}$) in the rat, *Am. J. Physiol.*, 209, 900–904, 1965.
22. Dunn, M.A., Green, M.H., and Leach, R.M., Jr., Kinetics of copper metabolism in rats: a compartmental model, *Am. J. Physiol.*, 261, E115–E125, 1991.
23. Dunn, M.A., Historical overview of copper kinetics, in *Kinetic Models of Trace Element and Mineral Metabolism During Development*, K.N.S. Subramanian and M.E. Wastney, Eds., CRC Press, Boca Raton, 1995, 171–185.

24. Bush, J.A., Mahoney, J.P., Markowitz, H., Gubler, C.J., Cartwright, G.E., and Wintrobe, M.M., Studies on copper metabolism. XVI. Radioactive copper studies in normal subjects and in patients with hepatolenticular degeneration, *J. Clin. Invest.*, 34, 1766–1778, 1955.
25. Bush, J.A., Mahoney, J.P., Gubler, C.J., Cartwright, G.E., and Wintrobe, M.M., Studies on copper metabolism. XXI. The transfer of radiocopper between erythrocytes and plasma, *J. Lab. Clin. Med.*, 47, 898–906, 1956.
26. Hazelrig, J. and Owen, C.A., Jr., A mathematical model for copper metabolism and its relation to Wilson's disease, *Am. J. Physiol.*, 211, 1075–1081, 1961.
27. Sternlieb, I. and Scheinberg, I.H., Radiocopper in diagnosing liver disease, *Sem. Nucl. Med.*, 2, 176–188, 1982.
28. Sternlieb, I., Gastrointestinal copper absorption in man, *Gastroenterology*, 52, 1038–1041, 1967.
29. Sternlieb, I., Morell, A.G., Tucker, W.D., Greene, M.W., and Scheinberg, I.H., The incorporation of copper into ceruloplasmin *in vivo*: studies with $copper^{64}$ and $copper^{67}$, *J. Clin. Invest.*, 40, 1834–1840, 1961.
30. Strickland, G.T., Beckner, W.M., Leu, M., and O'Reilly, S., Turnover studies of copper in homozygotes and heterozygotes for Wilson's disease and controls: isotope tracer studies with $^{67}Cu$, *Clin. Sci.*, 43, 605–615, 1972.
31. Strickland, G.T., Beckner, W.M., and Leu, M., Absorption of copper in homozygotes and heterozygotes for Wilson's disease and control: isotope tracer studies with $^{67}Cu$ and $^{64}Cu$, *Clin. Sci.*, 43, 617–625, 1972.
32. Janssens, A.R. and van den Hamer, J.A., Kinetics of $^{64}$Copper in primary biliary cirrhosis, *Hepatology*, 6, 822–827, 1982.
33. Vierling, J.M., Shrager, R., Rumble, W.F., Aamodt, R., Berman, M.D., and Jones, E.A., Incorporation of radiocopper into ceruloplasmin in normal subjects and patients with primary biliary cirrhosis and Wilson's disease, *Gastroenterology*, 74, 652–660, 1978.
34. Johnson, P.E., Lykken, G., Mahalko, J., Milne, D., Inman, L., and Sandstead, H.H., The effect of browned and unbrowned corn products on absorption of zinc, iron, and copper in humans, in *Maillard Reaction in Foods and Nutrition*, G.R. Waller and M.S. Feather, Eds., American Chemical Society, Washington, D.C., 1983, 349–360.
35. Lukaski, H.C., Lykken, G.I., and Klevay, L.M., Simultaneous determination of copper, iron, and zinc absorption using gamma ray spectroscopy: fat effects, *Nutr. Rep. Int.*, 33, 139–146, 1986.
36. Shields, W.R., Murphy, T.J., and Garner, E.L., Absolute isotopic abundance ratio and the atomic weight of a reference sample of copper, *J. Res. NBS*, 68A, 589–592, 1964.
37. Turnlund, J.R., Keyes, W.R., and Peiffer, G.L., Isotopic ratios of molybdenum determined by thermal ionization mass spectrometry for stable isotope studies of molybdenum metabolism in humans, *Anal. Chem.*, 65, 1717–1722, 1993.
38. King, J.C., Raynolds, W.L., and Margen, S., Absorption of stable isotopes of iron, copper, and zinc during oral contraceptive use, *Am. J. Clin. Nutr.*, 31, 1198–1203, 1978.
39. Ting, B.T.G., Kasper, L.J., Young, V.R., and Janghorbani, M., Copper absorption in healthy young men: studies with stable isotope $^{65}Cu$ and neutron activation analysis, *Nutr. Res.*, 4, 757–769, 1984.

40. Crews, H.M., Ducros, V., Eagles, J., Mellon, F.A., Kastenmayer, P., Luten, J.B., and McGaw, B.A., Mass spectrometric methods for studying nutrient mineral and trace element absorption and metabolism in humans using stable isotopes, *Analyst*, 119, 2491–2514, 1994.
41. Johnson, P.E., A mass spectrometric method for use of stable isotopes as tracers in studies of iron, zinc, and copper absorption in human subjects, *J. Nutr.*, 112, 1414–1424, 1982.
42. Hachey, D.L., Blias, J., and Klein, P.D., High precision isotopic ratio analysis of volatile metal chelates, *Anal. Chem.*, 552, 1131–1135, 1980.
43. Buckley, W.T., Huckin, S.N., and Budac, J.J., Mass spectrometric determination of a stable isotope tracer for copper in biological materials, *Anal. Chem.*, 54, 504–510, 1982.
44. Hachey, D.L., Wong, W.W., Boutton, T.W., and Klein, P.D., Isotope ratio measurements in nutrition and biomedical research, *Mass Spectrom. Rev.*, 6, 289–328, 1987.
45. Turnlund, J.R. and Keyes, W.R., Automated analysis of stable isotopes of zinc, copper, iron, calcium, and magnesium by thermal ionization mass spectrometry using double isotope dilution for tracer studies in humans, *J. Micronutrient Anal.*, 7, 117–145, 1990.
46. Turnlund, J.R., Zinc, copper, and iron nutrition studied with enriched stable isotopes, *Biol. Trace Element Res.*, 12, 247–257, 1987.
47. Turnlund, J.R., Stable isotopes of copper, molybdenum, and zinc used simultaneously for kinetic studies of their metabolism, in *Kinetic Models of Trace Element and Mineral Metabolism During Development*, K.N.S. Subramanian and M.E. Wastney, Eds., CRC Press, Boca Raton, 1995, 133–143.
48. Turnlund, J.R., Wada, L., King, J.C., Keyes, W.R., and Acord, L.L., Copper absorption in young men fed adequate and low zinc diets, *Biol. Trace Element Res.*, 17, 31–41, 1988.
49. Turnlund, J.R., Michel, M.C., Keyes, W.R., Schutz, Y., and Margen, S., Copper absorption in elderly men determined by using stable $^{65}Cu$, *Am. J. Clin. Nutr.*, 36, 587–591, 1982.
50. Turnlund, J.R., Keyes, W.R., Anderson, H.L., and Acord, L.L., Copper absorption and retention in young men at three levels of dietary copper by use of the stable isotope $^{65}Cu$, *Am. J. Clin. Nutr.*, 49, 870–878, 1989.
51. Turnlund, J.R., Keyes, W.R., Peiffer, G.L., and Scott, K.C., Copper absorption, excretion, and retention by young men consuming low dietary copper, determined using the stable isotope $^{65}Cu$, *Am. J. Clin. Nutr*, 67, 1219–1225, 1998.
52. Turnlund, J.R., Stable isotope studies of the effect of dietary copper on copper absorption and excretion, in *Copper Bioavailability and Metabolism*, Kies, C., Ed., Plenum Press, New York, 1990, 21–28.
53. Turnlund, J.R., Reager, R.D., and Costa, F., Iron and copper absorption in young and elderly men, *Nutr. Res.*, 8, 333–343, 1988.
54. Turnlund, J.R., Swanson, C.A., and King, J.C., Copper absorption and retention in pregnant women fed diets based on animal and plant proteins, *J. Nutr.*, 113, 2346–2352, 1983.
55. Ehrenkranz, R.A., Gettner, P.A., Nelli, C.M., Sherwonit, E.A., Williams, J.E., Ting, B. T. G., and Janghorbani, M., Zinc and copper nutritional studies in very low birthweight infants: Comparison of stable isotopic extrinsic tag and chemical balance methods, *Pediatr. Res.*, 26, 298–307, 1989.

56. Turnlund, J.R., King, J.C., Gong, B., Keyes, W.R., and Michel, M.C., A stable isotope study of copper absorption in young men: effect of phytate and alpha-cellulose, *Am. J. Clin. Nutr.*, 42, 18–23, 1985.
57. Knudsen, E., Sandstrom, B., and Solgaard, P., Zinc, copper and magnesium absorption from a fibre-rich diet, *J. Trace Elements Med. Biol.*, 10, 68–76, 1996.
58. Turnlund, J.R., Keyes, W.R., Hudson, C.A., Betschart, A.A., Kretsch, M.J., and Sauberlich, H.E., Zinc, copper, and iron absorption and retention by young women fed vitamin B-6 deficient diets, *Am. J. Clin. Nutr.*, 54, 1059–1064, 1991.
59. Taylor, C.E., Bacon, J.R., Aggett, P.J., and Bremner, I., Intestinal absorption and losses of copper measured using $^{65}Cu$ in zinc-deprived men, *Eur. J. Clin. Nutr.*, 45, 187–194, 1991.
60. Johnson, P.E., Milne, D.B., and Lykken, G., Effects of age and sex on copper absorption, biological half-life, and status in humans, *Am. J. Clin. Nutr.*, 56, 917–925, 1992.
61. Stuart, M.A. and Johnson, P.E., Intrinsic labeling of confinement-reared goslings with $^{65}Cu$ for use in human absorption studies, *Nutr. Res.*, 6, 203–213, 1986.
62. Johnson, P.E. and Lykken, G.I., $^{65}Cu$ absorption by men fed intrinsically and extrinsically labeled whole wheat bread, *J. Agr. Food Chem.*, 36, 537–540, 1988.
63. Johnson, P.E., Stuart, M.A., Hunt, J.R., Mullen, L., and Starks, T.L., $^{65}Cu$ absorption by women fed intrinsically and extrinsically labeled goose meat, goose liver, peanut butter and sunflower butter, *J. Nutr.*, 118, 1522–1528, 1988.
64. Weber, K.M., Boston, R.C., and Leaver, D.D., A kinetic model of copper metabolism in sheep, *Aust. J. Agric. Res.*, 31, 773–790, 1980.
65. Buckley, W.T., A kinetic model of copper metabolism in lactating dairy cows, *Can. J. Anim. Sci.*, 71, 155–166, 1991.
66. Buckley, W.T., Copper metabolism in dairy cows: development of a model based on a stable isotope tracer, in *Kinetic Models of Trace Element and Mineral Metabolism During Development*, K.N.S. Subramanian and M.E. Wastney, Eds., CRC Press, Boca Raton, FL, 1995, 37–51.
67. Scott, K.C. and Turnlund, J.R., Compartmental model of copper metabolism in adult men, *J. Nutr. Biochem.*, 5, 342–350, 1994.
68. Turnlund, J.R., Thompson, K.H., and Scott, K.C., Key Features of copper vs. molybdenum metabolism models in humans, in *Mathematical Modelling in Experimental Nutrition*, A.J. Clifford and H.-G. Muller, Eds., vol. 445, Plenum Press, New York, 1998, 271–282.
69. Levenson, C.W. and Janghorbani, M., Long-term measurement of organ copper turnover in rats by continuous feeding of a stable isotope, *Anal. Biochem.*, 221, 243–249, 1994.

# 9

# *Use of Stable Isotopes of Selenium to Investigate Selenium Status*

**Helen M. Crews**

**CONTENTS**

9.1 Introduction ... 130
9.2 Dietary Selenium and Its Metabolism ... 130
9.2.1 Sources and Daily Intakes ... 130
9.2.2 Chemical Form and Bioavailability ... 131
9.2.3 Metabolism of Selenium ... 132
9.3 The Role of Selenium in the Body ... 133
9.3.1 Selenium and Disease ... 133
9.3.1.1 Selenium Deficiency and Disease ... 133
9.3.1.2 Selenium and Cancer ... 134
9.3.2 Selenoproteins ... 134
9.3.2.1 Intracellular Glutathione Peroxidases (EC 1.11.1.9.) ... 135
9.3.2.1.1 Cellular (Cystolic) GSHpx ... 135
9.3.2.1.2 Phospholipid Hydroperoxide GSHpx ... 135
9.3.2.1.3 Gastrointestinal GSHpx ... 136
9.3.2.2 Extracellular GSHpx ... 136
9.3.2.2.1 Plasma GSHpx ... 136
9.3.2.3 Iodothyronine Deiodinases (EC 3.8.1.4.) ... 136
9.3.2.4 Thioredoxin Reductase (EC 1.6.4.5.) ... 136
9.3.2.5 Selenium-binding Protein ... 137
9.3.2.6 Others ... 137
9.4 Assessment of Selenium Status and Use of Stable Isotopes ... 137
9.4.1 Status Assays ... 137
9.4.2 Analytical Aspects ... 138
9.4.2.1 Assays for GSHpx Activity ... 138
9.4.2.2 Measurement of Selenium Isotopes ... 139
9.4.3 Modelling of Selenium Body Pools ... 140
9.4.4 Stable-isotope Studies with Low-to-medium Selenium Intakes ... 143
9.4.5 Stable-isotope Studies with High Selenium Intakes ... 144
9.5 Conclusion ... 145
References ... 146

0-8493-8730-2/00/$0.00+$.50

## 9.1 Introduction

Both interest in, and the number of publications concerning, selenium (Se) are vast and increasing. This remarkable element has only been recognized relatively recently as nutritionally essential, with biochemical function in animals first shown in 1973 by Rotruck et al.[1,2] Not only is selenium an essential element, but evidence of its toxic effect was also reported as long ago as the thirteenth century.[3] A third role postulated for Se is that of an anticancer agent and this has led to an even broader interest in its occurrence and utilization in the body.[3] The role of Se in human health is still not well understood. The complex, multifaceted nature of this element is influenced by the chemical form in which it occurs, as well as the physiological state and geographical location of the organism, which, in turn, can influence its uptake, absorption and bioavailability in humans and animals. Its behavior is intrinsically linked with that of other nutrients such as vitamin E and iodine. For an excellent summary and assessment of the literature pertaining to Se, Reilly's recent book is recommended to the reader.[3]

Thus, when assessing an individual's status with regard to Se, many factors may influence the accuracy of that assessment. Much of the groundbreaking work in understanding how Se functions has been undertaken with animal models, cell lines, and radioisotopic labelling. The latter can be ethically problematic for use with human volunteers and, in the last two decades, with the advent of better ways of measuring stable isotopes, more human studies at realistic dietary levels have been undertaken. The reader is referred to Chapter 1 for a discussion of advances in stable-isotope methodology and to Chapter 4 for methods of analysis for trace-element absorption. In this chapter, a brief overview of Se in relation to its occurrence in the diet and its role in the human body will be given first. Then the measures currently used to assess Se status will be introduced, followed by examples of work in which Se stable isotopes have been used to assess the status, or the factors influencing Se status, in humans. Finally, a summary of current capabilities and future requirements will be presented.

## 9.2 Dietary Selenium and Its Metabolism

### 9.2.1 Sources and Daily Intakes

Selenium in the human diet comes from a variety of sources. For example, data from the 1994 and 1997 U.K. total diet studies show that offal (0.42 and 0.49 mg/kg for 1994 and 1997, respectively), fish (0.39 and 0.36 mg/kg), nuts

(0.29 and 0.25 mg/kg), and eggs (0.19 mg/kg for both 1994 and 1997) contain the highest mean concentrations of Se; however, dietary exposure (μg/day) is greatest from meat products (6), bread (5), fish (5), miscellaneous cereals (4), poultry (4) and milk (4).[4,5] These food groups provide approximately 70% of the U.K. daily intake of 40 μg/day.[4,5] This intake compares with that found, for example, in Belgium[6] of 55 μg/day, in Finland of 113 μg/day, and in the U.S. of 98 μg/day.[6–8] In the People's Republic of China, where there are well characterized areas of geographically distinct high and low Se occurrence, Yang et al. reported intakes ranging from 3 to 11 μg/day in the Keshan area (hence the name given to a Se responsive cardiomyopathy — Keshan disease) to 3200 to 6690 μg/day in Enshi Province.[9] Later, the same group suggested a marginal daily safe Se intake of 750 to 850 μg/day based on studies with residents of Enshi Province.[10] Recently, Janghorbani et al. reported results from a study of Chinese male residents of Jianshi County in Hubei Province with estimated long-term dietary intakes of 197 to 1230 μg/day.[11] In contrast, men and women from the low Se area of South Island, New Zealand, may have intakes in the range of 20 to 30 μg/day without suffering deficiency problems.[12,13]

It is apparent that very large differences in Se intake exist worldwide and the extremes of dietary exposure may reflect the soil content of Se in an area. In addition, the import and export of foods between countries influence access to a diet which contains Se. Thus, Golubkina and Alfthan reported that in 27 regions of Russia, three distinct groups of the population with different serum Se values were found; these were determined primarily by the use of wheat which was high (American or Australian origin) or low (domestic or European origin) in Se content.[14] In a region of endemic low Se, Khabarovsk, an unexpectedly high Se status was found due to the use of imported high Se wheat. However, in the Irkutsk region, which was reported to have high Se concentrations in local spring waters, domestically produced wheat was consumed and low serum Se values were found.[14]

### 9.2.2 Chemical Form and Bioavailability

The chemical forms of Se most commonly referred to in dietary studies are the inorganic forms selenate ($SeO_4^{2-}$) and selenite ($SeO_3^{2-}$), and the organic forms selenomethionine ($CH_3SeCH_2CH_2CH(NH_2)CO_2H$) and selenocysteine ($HSeCH_2CH(NH_2)CO_2H$). The chemical form of Se may influence the amount of the element absorbed from the gastrointestinal tract. Generally more Se (95 to 97%) is absorbed as selenomethionine (SeMet) than when it is present as selenite (44 to 70%).[15] However, earlier animal studies indicated that when Se was present as selenocysteine (SeCys), it was retained in the body in a similar fashion to selenite.[16] Studies with rats have demonstrated that SeMet is absorbed by an active transport mechanism shared with methionine, while selenite is absorbed by simple diffusion, selenate by sodium mediated carrier transport shared with sulphate and SeCys may share a common active transport mechanism with basic amino acids.[17–19] The bioavailability of Se from the diet, i.e., the fraction of ingested Se that is utilized for normal physiological functions or storage, is thus primarily determined by the form of Se ingested and not by the amount absorbed.[20]

The information on the form of Se in foods is erratic and often conflicting, with research on the form of Se in yeast producing much interest since it is frequently used in intervention studies and is found in some dietary supplements.[19] Broadly speaking, foods of animal origin contain SeCys and those of plant origin SeMet. However, because these organisms also metabolize Se to a greater or lesser extent, inorganic Se and other organic forms of the element will exist in foodstuffs. Human studies of bioavailability of different chemical forms of Se in foodstuffs can be done using stable-isotope labelling.[21] This permits the Se species to be intrinsically labelled in the plant or animal, and aids identification of the metabolic fate of the absorbed element, which can be characterized using changes in status and kinetic modelling.[18,21] These approaches will be discussed in Section 9.4.

### 9.2.3 Metabolism of Selenium

In biological systems, selenate is reduced to selenite, by enzymatic activation with adenosine triphosphate sulphurlyase in the presence of magnesium ions to form adenosine-5′-selenophosphate, followed by non-enzymatic cleavage using reduced glutathione (GSH).[22] Selenite is reduced further via GSH, reduced nicotinamide adenine dinucleotide phosphate (NADPH), and glutathione reductase, to hydrogen selenide ($H_2Se$).[23] Selenide ($Se^{2-}$) is considered to be the key form of Se during metabolism.[22,23] It is, however, one of the most toxic forms of Se, and detoxification of $Se^{2-}$ by successive methylation produces monomethylselenol ($CH_3SeH$), dimethylselenide ($(CH_3)_2Se$), and the trimethylselenonium ion ($(CH_3)_3Se^+$). The production of these methylated forms is increased with higher dietary intakes of Se.[11,23]

SeMet and SeCys are treated in different ways by the body. SeMet is bound non-specifically to hemoglobin in red blood cells and to albumin in plasma for transport around the body. A percentage of ingested SeMet follows the metabolic pathways of methionine so that much (approximately 70%) of this Se undergoes non-specific incorporation into proteins in muscle tissue while the remainder goes to the liver.[23] Some SeMet may be catabolized to selenite after absorption.[22] There is evidence that SeCys can substitute for cysteine at high concentrations, but, generally, the Se from SeCys is considered to be metabolized in a similar fashion to selenite.[16,23,24] Decomposition of SeCys to reduced Se is catalyzed by the enzyme selenocysteine β-lyase.[23]

The metabolism of Se is still not well understood and the role of molecular biology, particularly with reference to selenoprotein synthesis, has only emerged in the last 10 to 15 years.[23,26] The biosynthesis of SeCys for specific incorporation, by co-translation, into the active sites of a number of proteins has so far been most comprehensively described for prokaryotic selenoproteins.[26] Figure 9.1 is a simplified overview of the current understanding of the way in which dietary Se is metabolized by the body.

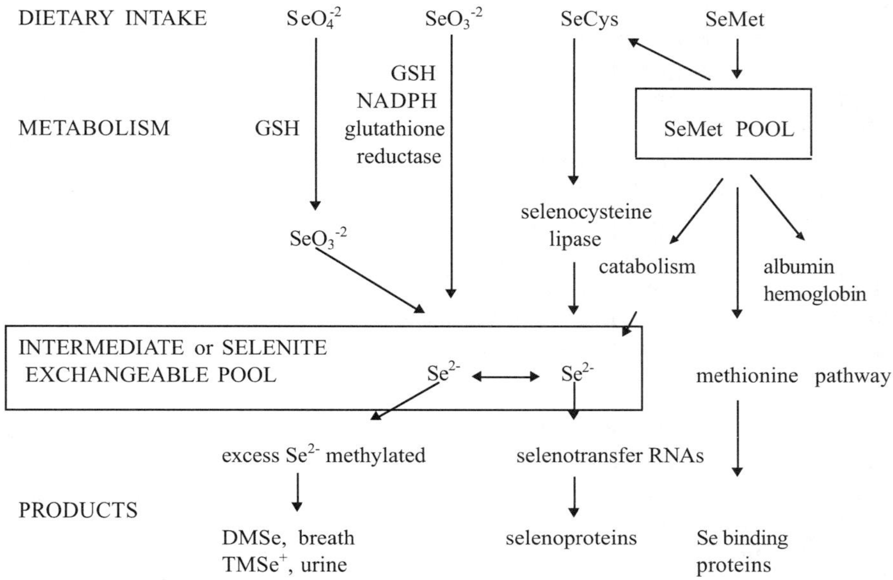


**FIGURE 9.1**
Simplified overview of the metabolic fate of dietary Se. (Adapted from References 11, 23, and 24.)

## 9.3 The Role of Selenium in the Body

### 9.3.1 Selenium and Disease

#### *9.3.1.1 Selenium Deficiency and Disease*

Since the recognition of the essentialness of Se, decreased supplies of Se have been associated with a variety of clinical conditions.[27] These include cardiomyopathy (Keshan disease), which affects young children and women of child-bearing age, osteoarthropathy (Kashin-Beck disease), which usually develops in children age 5 to 13 years, and possibly ischemic heart disease.[28,29]

Recently, the intriguing role of Se in relation to viral disease and cardiomyopathy has been discussed by Beck.[30] Although supplementation with Se prevents the occurrence of Keshan disease, the disease has a seasonal and annual incidence and not all individuals deficient in Se develop the disease. Coxsackieviruses have been isolated from the blood and tissue of Keshan disease victims and Beck used a mouse model to study the relationship between the virus and Se deficiency. Infected by two strains of the virus, one that causes myocarditis (CVB3/20) in mice and one that does not (CVB3/0) — even though both occur in heart tissue, mice that were Se-deficient were compared with Se-adequate mice. Those mice deficient in Se developed more severe myocarditis when exposed to CVB3/20 than did the replete mice. In addition, Se-deficient mice exposed to the benign virus developed moderate myocarditis, whereas the Se-adequate mice did not. It was therefore concluded that the low Se status altered the virulence of the normally benign CVB3/0 virus and subsequent studies demonstrated that the benign virus itself had changed due to replication in a Se-deficient host.[30]

It can be dangerous to make cross-species comparisons; it is clear, for example, that mice and humans do not respond to Se deficiency or excess in the same way.[31] Nevertheless, the fact that the *nutritional status of the host organism* (including an assessment of vitamin E status when considering the behavior of Se) altered the virulence of a viral pathogen by changing the phenotype of the virus implies that this effect probably occurs in humans as well.[30,31]

#### 9.3.1.2 Selenium and Cancer

Studies with rodents indicate that Se supplementation at levels above dietary requirements is capable of lowering the incidence of tumorigenesis induced by chemical carcinogens or viruses.[32,33] A recent human study indicated prevention of certain cancers when daily supplementation with yeast containing Se at 200 μg was given.[34] This level is above both the United Kingdom Reference Nutrient Intake (70 and 60 μg/day for men and women, respectively) and the United States National Research Council Recommended Daily Allowance (70 and 55 μg/day for men and women, respectively).[35,36] In order to understand how best to provide dietary sources for both essential and disease preventative functions, our knowledge of, and ability to measure, biomarkers of Se status need some clarification.

### 9.3.2 Selenoproteins

Selenium occurs in tissues associated with proteins, both loosely bound and as Se analogues of sulphur amino acids.[37] Selenoproteins contain SeCys at their active site and are Se-dependent to differing degrees, i.e., there is differential regulation of selenoproteins. Replacement of the Se with less active sulphur (to give cysteine) reduces activity.[26] In contrast, levels of Se-binding proteins are not apparently regulated by the availability of Se[27] and catabolism releases Se.

Some thirty selenoproteins have been identified in mammalian tissues by use of *in vivo* labelling with the radioisotope $^{75}$Se and many have been further characterized by purification and/or cloning.[38] However the functions of only a few are beginning to be known. Some of the best understood will be briefly described in the next sections since they will be the ones most likely to be used as current markers of functional status; knowledge of their behavior is necessary when attempting to interpret human studies using Se stable isotopes.

Details concerning the structure of these proteins are given in the recent review by Patching and Gardiner and are not repeated here.[26] The EC number following some of the names is that given by the Commission on Enzymes of the International Union of Biochemistry.[23]

#### 9.3.2.1 Intracellular Glutathione Peroxidases (EC 1.11.1.9.)

The intracellular glutathione peroxidases (GSHpx) have been well described by Sunde and are summarized in the following sections.[39]

##### 9.3.2.1.1 *Cellular (Cytosolic) GSHpx*

Classical cellular or cytosolic GSHpx was discovered by Mills and also investigated by Rotruck et al. when looking for an enzymatic function to explain the antioxidant activity of Se.[2,40] Cellular GSHpx appears to function in conjunction with vitamin E and is located in the cytosol and mitochondrial matrix, whereas vitamin E is present in the cell membranes.[37] One function is to provide protection of erythrocytes (red blood cells) against oxidative hemolysis, but the complete role of cellular GSHpx is still not clear. Since in Se-deficient animals cellular GSHpx mRNA is very low, and any free Se could be diverted to other selenoproteins whose mRNA levels are not significantly reduced during Se deficiency, Sunde proposes that cellular GSHpx in the liver acts as a homeostatic mechanism.[39] It keeps free Se levels low and ensures that it is diverted to the most important functions in times of Se deficiency, such as those involving phospholipid hydroperoxide GSHpx, plasma selenoprotein P, or Se-dependent 5′-deiodinase. The hypothesis, supported by work with Se-deficient animals, is that Se is released and goes to the three latter proteins and then, as intracellular Se increases, the cellular GSHpx mRNA stabilizes and Se is able to be incorporated into cellular GSHpx. This would reduce the level of $Se^{2-}$, which is toxic to the cell. Thus GSHpx in the liver buffers the internal environment. Sunde suggests that cellular GSHpx mRNA regulation by Se status is an evolutionary-conserved mechanism that all cells use to sense Se status and to control Se metabolism.[39] Therefore, cellular GSHpx is considered to have both an antioxidant and a Se-storage function.[27]

The two most widely used measures of human Se status are the concentration of the element in whole blood or its fractions and the activity of GSHpx in erythrocytes.[37] However, as Diplock reported in 1993, having taken into account all the human studies available at that time, there was a good correlation between blood or plasma Se concentration and GSHpx activity in erythrocytes at blood concentrations of up to about 1 μmol Se/L (79 mg/L).[32] Above this level the activity became saturated.

##### 9.3.2.1.2 *Phospholipid Hydroperoxide GSHpx*

Phospholipid hydroperoxide GSHpx acts at the interface between membranes and the aqueous phase of the cell and occurs in the same organs as cellular GSHpx. It metabolizes phospholipid hydroperoxides.[37] It may be an important enzyme for heart protection (c.f., Keshan disease in severe Se deficiency) and is not regulated by Se status in the same way as cellular GSHpx.[39]

*9.3.2.1.3 Gastrointestinal GSHpx*

The role of gastrointestinal GSHpx is not well understood. So far, mRNA for this protein has been found only in rat gastrointestinal cells; other workers have reported it to be synthesized mainly in the human liver and colon.[39,41]

#### 9.3.2.2 Extracellular GSHpx

*9.3.2.2.1 Plasma GSHpx*

There is a considerable amount of information about the structure of plasma GSHpx (which is similar to that of the cellular form), but little is known about its function.[27] The kidneys and lungs are thought to be the main sites of synthesis.[27,37,42] The enzyme is found not only in plasma, but also in breast milk and lung fluids.[42] The concentration of GSH in plasma is very low and this has led to postulations that the function of this enzyme may not be that of a glutathione-dependent, lipid hyperoxide reducing enzyme in plasma.[27,42] The levels of GSH in the kidney would allow the enzyme to act as a peroxidase and, thus, it may have a specific function in kidney tubules.[27,37] It does, however, appear to be a good indicator of Se status, despite its function not being understood.[42] In human subjects of low Se status, the levels of dietary Se required to maximize plasma GSHpx activity were thought to represent a nutritional requirement based on physiological function.[38,43]

#### 9.3.2.3 Iodothyronine Deiodinases (EC 3.8.1.4.)

Iodothyronine deiodinases (IDs) selenoproteins control the conversion of inactive plasma thyroxin (T4) to biologically active plasma 3,3′,5-triiodothryronine (T3) and further inactive metabolites such as 3,3′-diiodothronine (T2).[27,38,44] Three distinct but closely related IDs have been identified and are termed Types I (IDI), II (IDII), and III (IDIII).[26] All contain SeCys at their active site. IDI is responsible for the conversion of T4 to T3, and shares this role with IDII, which, although not a selenoenzyme, is adversely affected by Se deficiency.[23] IDIII promotes the degradation of T3 to T2, as well as the 5-deiodination of T4 to produce reverse T3 (3,3′,5′-triiodothyronine) which appears to be inactive but which may inhibit the action of T3.[26]

#### 9.3.2.4 Thioredoxin Reductase (EC 1.6.4.5.)

Thioredoxin reductase has been recently discovered and catalyzes the NADPH-dependent reduction of the redox protein thioredoxin and, with this protein, it is responsible for the activation and deactivation of some transcription factors and ribonucleotide reductase, which is essential for DNA synthesis.[37,45,46] It has been suggested that the reduction of thioredoxin by thioredoxin reductase is important for the growth of both normal cells and cancer cells, making Se deficiency a risk factor for cancer development.[26]

#### 9.3.2.5 Selenium-binding Protein

In humans and rodents, 60 to 80% of the Se in plasma occurs as the extracellular protein, selenoprotein P (SeP), which contains 5 to 10% of the total body Se.[22,47] As with most of the selenoproteins, SeP has been characterized in some detail, but its exact function is still unknown. Under conditions of limited Se availability, it has been shown that SeP synthesis takes precedence over the synthesis of GSHpx and that the half-life of SeP is approximately one-third that of GSHpx (both half-lives unaffected by Se status).[48] The amino acid sequence of SeP is highly conserved between the rat and man.[48] The numerous histidine and cysteine residues may be suitable for binding free transition metals, which might enable SeP to act as a metal detoxification site.[26] It has also been speculated that SeP is a transport protein, but the presence of Se in its primary tissue, and the subsequent waste of the energy used to incorporate the Se if it had then to be converted back to inorganic Se for synthesis of other selenoproteins, argue against this.[27,47] Appearance of SeP coincided with Se-produced protection against diquat-induced liver injury and lipid peroxidation; it is thus thought to have antioxidant properties, but this remains uncertain.[47]

#### 9.3.2.6 Others

Sperm capsule selenoprotein is found in the mitochondrial capsules.[26] It is thought to have a structural role in maintaining the tail structure of spermatozoa since this selenoprotein contains several intramolecular sulphydryl bonds which would confer stability. This role would explain the abnormal sperm development found in the rat under conditions of Se deficiency.[27]

Selenoprotein W is low weight protein which has been isolated from rat and human muscle but its function is not known.[26,27]

Two selenium-binding proteins, one at 14 KDa and another at 56 KDa, have been studied. The former is a fatty-acid-binding protein and the latter is more closely related to drug-binding proteins.[23] The levels of these proteins are not apparently regulated by Se availability and more work is required to ascertain their functions.[27]

## 9.4 Assessment of Selenium Status and Use of Stable Isotopes

### 9.4.1 Status Assays

Measures of status are used to assess the nutritional adequacy of the diet and usually involve the concentration of the nutrient in biological tissues, functional tests, and biochemical assays.[49] In addition, clinical observations have been developed to assess Se status in various conditions ranging from deficiency to excessive intake.[50] Ideally, required now are measures of "optimal" status to allow an assessment of nutritional and disease (those diseases not due to severe Se deficiency) preventative status.

Nève and Carpentier recommended in 1989 that Se status could be assessed by: (1) plasma Se, which was sensitive to changes in Se status in humans, particularly at low-to-moderate intakes; (2) erythrocyte GSHpx activity, which offered an easily accessible functional index, although it was not sensitive to modification; (3) plasma GSHpx which was more sensitive to increasing status in severely depleted subjects; and (4) platelet GSHpx which offered better sensitivity to status modifications.[50]

In 1993, Diplock concluded that the preferred indices of human Se status were blood, or plasma and/or serum concentrations of the element, and the level of activity of GSHpx in erythrocytes or plasma, subject to certain caveats.[32] Two factors were considered to play an important part in forming a judgment about the reliability of blood or other tissue measurements of Se as an index of Se status: (1) the methodology for making the measurement of Se had to be reliable and reproducible, and (2) the variable measured had to be directly related to the biochemical variables for Se activity *in vivo*.

By the time Diplock's paper was published, a few stable-isotope studies with Se had been published. How much have these and subsequent stable-isotope studies helped in determining Se status in humans?

### 9.4.2 Analytical Aspects

#### *9.4.2.1 Assays for GSHpx Activity*

From an analytical viewpoint, the methods for some functional-effect assays, (GSH peroxidase activity in whole blood, serum, plasma, and platelets) are less robust than those for total Se or Se isotope determinations.

It should be noted that the methods for functional assays are generally not standardized nor collaboratively tested, and variants of two generally accepted methods are used.[51–53] The enzymic techniques for measurement of GSHpx activity can be difficult, with limited sensitivity, specificity, and stability, as well as problems in obtaining a constant blank and defining a reference method.[54–56] Some radioimmunoassays have been developed for GSHpx serum and plasma, for example.[56,57] Commercial kits based on radioimmunoassay methods are available, but these can be expensive, especially when considering population studies.

In addition, functional-effect assays using GSH have some limitations in their use. For example, erythrocyte GSH activity represents an easily accessible functional index of Se status, although it offers poor sensitivity to modification. Increased activity was slow after supplementation of depleted subjects with 100 to 200 μg Se/day, reflecting the long life span of red blood cells in circulation.[50] In addition, platelet GSHpx activity can reflect recent changes in intake and body stores rather than long-term status.[55] Also, for GSHpx activity in plasma or blood cells there appears to be a plateau for the

Se concentration in blood or plasma, beyond which there is no increase in activity.[58] For erythrocyte GSH, the peroxidase activity plateaus at plasma Se levels between 0.76 and 0.89 μmol/l and in whole blood at 0.76 and 2.03 μmol/l in different studies;[49,58] for platelet GSH activity, the plasma Se plateau has been found to be 1.39 to 1.71 μmol/l and 1.25 to 1.45 μmol/l.[59,60] In a recent human dietary intervention study, erythrocyte and plasma GSH peroxidase were found not to be sensitive to large changes in Se intake over a six-week period. Plasma Se concentrations, determined after intakes of 25 and 425 μg/day for six weeks, (mean plasma Se 64.6 and 103.8 μg/l, respectively) and platelet GSH peroxidase activity (for 425 μg/day intake, mean plasma Se 103.8 μg/l) were more sensitive indicators of change in Se status.[61]

#### 9.4.2.2 Measurement of Selenium Isotopes

Methods for determining total Se levels are generally well established and the number of certified reference materials now available should aid the provision of reliable data.[51] In addition, interlaboratory proficiency testing schemes are increasingly undertaken. Most methods of determination, including inductively coupled plasma-mass spectrometry (ICP-MS) but excluding purely instrumental methods such as neutron activation and X-ray fluorescence analysis, require the sample to be a dissolved mineralized state.[62] For human tissues, wet ashing with acid (nitric and/or sulphuric and/or perchloric) is preferred. Extreme care and strictly controlled procedures are required to prevent losses of Se.[63]

For determination of the individual isotopes of Se (isotope and % natural abundance: $^{74}$Se, 0.89; $^{76}$Se, 9.36; $^{77}$Se, 7.63; $^{78}$Se, 23.78; $^{80}$Se, 49.61; and $^{82}$Se, 8.73), a mass separation is needed, which means that mass spectrometric methods are required. The basic sample preparation for isotope measurements is not dissimilar to that for total Se determinations, and natural isotopic-abundance-certified reference materials or in-house reference materials can be used to assess any mass bias of an instrument. Techniques such as gas chromatography-mass spectrometry (GC-MS), which is widely available, and the increasingly available ICP-MS must be used. Because it does not require sample derivatization like GC-MS, and because of its high sample throughput, ICP-MS has become popular for isotopic studies.[64–68] The quadrupole instruments, with unit mass resolution, have a major limitation in that the dimer from the ICP's argon plasma, $^{40}Ar_2$, has the same mass as the most abundant Se isotope, $^{80}$Se, precluding measurement of this isotope. However, it is possible to accurately calculate the abundance of this isotope by deconvolution techniques if the other five are determined.[69,70] (See Chapter 3 for the mathematical treatment of stable-isotope data.) The use of a nitrogen microwave-induced ionizing plasma was used to obtain $^{80}$Se/$^{78}$Se ratios with a precision of 0.5% relative standard deviation (RSD) for standard solutions of 100 ng/ml and was applied to isotope dilution analysis of biological reference materials.[71] It is hoped that in the future, recently developed ICP-MS instruments, with collision or reaction cells situated between the plasma source and the mass detector, will enable the removal of the argon interference. The use of hydride generation with ICP-MS can minimize the interference from chloride ($^{40}Ar^{37}Cl$) in the sample or re-agents on the $^{77}Se$ signal.[67,68,72,73] Reference materials can help in the assessment of the accuracy of natural isotope ratio measurements as well as the use of standard additions and isotope dilution analysis.[67,71,73,74]

The biggest challenge in measuring individual stable isotopes with altered abundances (for both nutritional and geological studies) is to achieve the necessary accuracy and precision for the isotope ratios when one isotope may be present in excess compared with others. For GC-MS, precisions for Se isotope ratio measurements using isotope dilution were in the range 1 to 7% RSD using *o*-phenylenediamines as chelating agents.[65,66,75] For measurements by ICP-MS, the precision for isotope ratios for plasma, urine and fecal samples is in the range 0.1 to 1.0% RSD, and can vary with instrument type and age.[73,76] The achieveable precision is limited by Poisson ion-counting statistics for the instruments, as well as memory effects in the ample introduction systems for instruments.[58,72,73] Sample matrix and size, as well as Se concentration for each isotope, also influence precision and, while the isotope doses should be small enough not to perturb steady-state conditions in the body, the detection limit of the analytical technique used to determine the isotopic enrichment must be taken into account when calculating doses.[77] However, values of 1% or better are generally adequate for use in nutritional and clinical studies if Se concentrations are optimized as far as possible by judicious use of isotope labels. (See Chapters 1 and 4 for discussions of study design.) Precision values of 0.5% or better enable the data to be used for kinetic modelling of Se body-pool sizes, which is considered in the next section.

### 9.4.3 Modelling of Selenium Body Pools

Discussions of compartmental modelling and turnover of metabolic pools are given in Chapters 3 and 7, respectively. This type of study requires some form of marker and is a good example of Se stable isotopes bringing additional scope for improving the knowledge base when compared to or combined with other measures of Se status. Stable isotopes do not have the ethical problems of radioisotopes, but it is important to realize that straightforward analogies between radio- and stable-isotope tracer data cannot be made. The fact that stable isotopes have a mass which is not negligible, and also have considerable background levels in the body, means that direct comparisons between the two approaches will provide incorrect results.[70] Briefly, stable isotopes can be employed to perform compartmental analysis and kinetic modelling, following introduction into the body of the enriched isotope by oral and/or intravenous means, and the rate of appearance/disappearance in plasma and/or urine and/or feces measured. Simple kinetic models can be developed from a knowledge of the element's metabolism, and rate constants can be estimated and entered into the model. Using differential equations, the correct rate constants have been selected when the estimated isotope appearance/disappearance is identical to the measured data.[70] In a recent paper by Wilson and Dainty, the point is made, using Se as an example, that, in discussions about stable-isotope labels, the enriched isotope is often referred to as the tracer when, strictly speaking, the tracer is the whole isotopic spectrum.[78] This can be a confusing concept, but if enriched $^{82}Se$ were used to alter the isotopic spectrum, then when the $^{82}Se$ is measured in feces, for example, it will contain $^{82}Se$ from all sources of the element, not just the enriched isotope.[78]

Work by Janghorbani and colleagues introduced the use of stable isotopes for intrinsic tagging of some foods and for the quantitative determination of the component of body-pool Se that they defined as the selenite ($SeO_3^{2-}$) exchangeable pool.[79–82] Figure 9.1 shows the two body pools commonly referred to by Janghorbani and others.[8,11,23,24] The intermediate or selenite exchangeable-pool contains those compounds that can be derived from $SeO_3^{2-}$, while the SeMet pool holds SeMet-containing proteins.

In the 1990 study, Janghorbani et al. used the well established principles of *in vivo* stable-isotope dilution as applied extensively for the measurement of protein metabolism, total-body water, carbohydrate and fat metabolism.[82,83–85] However, the authors pointed out that the use of a stable isotope of Se and the greater degree of complexity of body Se compartments were major differences the single compartment models used for exchangeable electrolytes.[82] This important paper deserves some detailed discussion since the principles applied in it, and its results, are frequently cited in subsequent publications by Janghorbani and colleagues and by other workers.

In two experiments, healthy North American male volunteers had either a self-selected diet (two volunteers) plus a single dose (100 μg Se) of $^{74}SeO_3^{2-}$ or a controlled low-level Se diet (four volunteers) plus ten days of 100 μg Se as $^{74}SeO_3^{2-}$ as a dietary supplement. In the third experiment with four volunteers for each stage, $^{74}SeO_3^{2-}$ was administered in four ways; (1) a single oral dose of 82 μg after 25 days on a low basal diet supplemented with unlabelled $SeO_3^{2-}$ to provide a total of 120 μg Se each day; (2) as just described but with the $^{74}SeO_3^{2-}$ administered intravenously over a 4-hour period; (3) as in step 2 but no Se supplementation so that daily intake was 20 μg Se; and (4) as in step 3 but with the $^{74}SeO_3^{2-}$ administered intravenously as in 2. The $^{74}Se/^{76}Se$ ratio was measured (by NAA and by ICP-MS) in plasma and urine samples of the subjects for 13 days after receiving the $^{74}SeO_3^{2-}$ dose. Statistically, the urine vs plasma ratios were not significantly different. To determine the size of the selenite exchangeable pool, $W_{Se\text{-}EMP}$, at time t after administration, Janghorbani et al. used the following equation:[82]

$$W_{Se\text{-}EMP} = [k_1.{}^aSe^*_{r.t}.(1-k_2.R_{a/b,t})]/(R_{a/b,t}-R_{a/b}^0),$$

where $W_{Se\text{-}EMP}$ is the measured size of the pool at time t after the administration of the $^{74}SeO_3^{2-}$ dose and reflects a pool with average isotopic composition identical to the sampling compartment (in mg); ${}^aSe^*_{r.t}$ is the amount of isotope "a" ($^{74}Se$) retained by the body at time t after administration of the label (*denoted originating from the labelled solution); $R_{a/b,t}$ is the isotope ratio (wt.:wt.) for isotope "a" to isotope "b" ($^{76}Se$) determined at time t after administration of the label; $R_{a/b}{}^0$ is the baseline isotope ratio in the sampling compartment (plasma or urine); $k_1$ is the natural ratio of $^aSe$ to $^bSe$; and $k_2$ is the ratio of $^aSe$ to $^bSe$ in the administered labelled $SeO_3{}^{2-}$ solution.

Using this calculation, it was found that the exchangeable $SeO_3{}^{2-}$ pool correlated positively with daily Se intake in the volunteers consuming diets of both known and variable content, decreasing on a Se-restricted diet. The route of administration of the isotope had no effect on the pool size. The authors concluded that measurement of $W_{Se\text{-}EMP}$ could prove to be an acceptable index of Se status and that it had distinct advantages over enzyme assays.[82] The size of the pool was not expected to be dependent upon many physiological variables that were known to influence the turnover of proteins by which enzyme assays for status were limited; $W_{Se\text{-}EMP}$, reflected the whole-body pool in contrast to plasma Se determinations which may have no bearing on intracellular concentrations of the element. The authors accepted that the approach required further development, particularly in relation to the quantitative size of $W_{Se\text{-}EMP}$ and the whole-body pool, and the effect of SeMet degradation on influx into $W_{Se\text{-}EMP}$. In addition, $W_{Se\text{-}EMP}$ needed to be related to functional indices of Se over a wide range of dietary exposures.

In 1999, an intervention study was completed in the U.K. by Fox et al. in which indices of Se status in humans were studied by measurements of the $SeO_3{}^{2-}$ exchangeable pool and the plasma pool, as well as measurements of platelet, plasma, and erythrocyte GSHpx activities.[61] Twelve healthy adult males were given three different diets of medium (75 μg Se/day to reflect European intakes), low (25 μg Se/day to reflect low Se intakes), and high (425 μg Se/day to reflect supplemented diets) Se levels for a period of six weeks each. The low basal diet was supplemented with Se-enriched brewer's yeast to attain the medium and high intakes. During the sixth week of each intervention period, each volunteer was given an intravenous dose of 88 μg Se as $^{74}SeO_3{}^{2-}$ via an indwelling cannula and blood samples taken at half-hour intervals for the first four hours and then hourly for the next four hours. Further blood samples were taken by venupuncture at 24, 48, 72, and 168 hours after the dose. Plasma Se concentration and Se isotopes (except $^{80}Se$ which was calculated by deconvolution) were measured using hydride generation with ICP-MS. Data from the plasma samples were analyzed using the two-compartment model with SAAM II software (SAAM Institute Inc., Seattle, WA, U.S.A.) to calculate the rate constants for the movement between the compartments and loss from the system. GSHpx activity was measured using an automated version of the method of Paglia and Valentine.[52,61]

The estimated exchangeable body-pool sizes changed significantly as a result of differences in Se intake and correlated well with changes in plasma Se concentrations (R- 0.833, $p<0.001$).[61] This result demonstrated the usefulness of plasma Se concentration as an index of body levels of exchangeable Se. The study also demonstrated that consumption of a high diet, which contained 400 μg Se as Se from yeast, increased the exchangeable body pool of

Se, presumably in non-specific protein tissue which could accumulate over time, the effects of which are not known. The GSHpx assay results showed that there were no changes in activity in erythrocyte and plasma samples, but there was a significant increase in platelet activity following the high Se diet and a decrease after the medium Se diet, but no change after the low diet.[61]

An earlier publication by Swanson et al. had applied stable-isotope methods to the utilization of Se by pregnant and non-pregnant women.[86] Apparent absorption from a defined diet of 150 μg Se/day was 80% for women in early and late pregnancy as well as for non-pregnant controls. Pregnant women tended to conserve Se by decreasing urinary excretion. These findings were corroborated by monitoring urinary and fecal excretion of 40 μg Se as $^{76}Se$ in intrinsically labelled egg. The isotope data also indicated that recent Se intake was incorporated into a long-term Se pool. Mean GSHpx activity was lower in plasma and higher in platelets in the pregnant women as compared with controls, but the physiological significance was not understood.

### 9.4.4 Stable-isotope Studies with Low-to-medium Selenium Intakes

For the purposes of this discussion, low-to-medium intakes will be those intakes up to about 100 μg Se/day. Higher intakes will be considered when data are compared within an experiment. Finley, in 1999, investigated the retention and distribution of stable isotopes in healthy male volunteers after consumption of $^{74}SeO_4^{2-}$ or $^{74}SeO_3^{2-}$ or broccoli intrinsically labelled with $^{82}Se$.[87] The stable isotopes were consumed in a test meal after consuming either low (32.6 μg Se/day) or high (226.5 μg Se/day) Se diets for 85 days. Urine, fecal, and blood samples were collected for the remainder of the 105-day study. Isotope absorption was not different for the $^{74}SeO_4^{2-}$ and $^{82}Se$ broccoli. Absorption of $^{74}SeO_3^{2-}$ was very variable and not included in statistical analyses. More isotope was retained by those on the higher diet, and broccoli $^{82}Se$ was better retained than that from $^{74}SeO_4^{2-}$. However, plasma Se contained less broccoli $^{82}Se$ than $^{74}SeO_4^{2-}$. Less isotope was found in plasma proteins of subjects fed the high Se diet, but the form of Se had no effect on the isotope distribution in the plasma.

In a another recent publication by Finley and colleagues, healthy men and women living in an area of low Se intake on the South Island of New Zealand were given daily supplements (0–40 μg Se/day) to compare retention of stable Se before and after consumption of supplements.[13] The hypothesis was that if Se supplementation changed the Se status or the need for Se by the subjects, such a change would be reflected by changes in retention of stable isotopes of Se. The women (29) and men (15) were given 100 μg $^{74}Se$ as $^{74}SeO_4^{2-}$ in water after an overnight fast. Blood was collected during the next three weeks. After this period, the subjects were divided into five groups and consumed daily, for six months, a tablet containing 0, 10, 20, 30, or 40 μg Se as L-SeMet. They were given a second dose of 100 μg $^{74}Se$ as $^{74}SeO_4^{2-}$ during the last three weeks of the supplementation period. The conclusion from the experiment was that Se supplementation had relatively little effect on the retention of stable $^{74}Se$, especially tissues that reflect long-term intake, i.e., erythrocytes and platelets. Retention by the plasma, which reflects short-term uptake, was decreased by supplemental plasma but the decrease was small. It was postulated that the subjects had adapted to their long-term low Se intakes by increased retention of Se. Reduced urinary excretion may offer a partial explanation for the increase. Another adaptation may be related to the distribution of Se among plasma proteins. Plasma was chromatographic to separate out plasma proteins for the 0 and 30 μg Se/day groups. For the 30 μg Se/day group, increased $^{74}Se$ retention in the plasma was accompanied by a significant (P=0.005) decrease in a peak in the plasma considered to be SeP[87] and no change in a second peak thought to be albumin.[87] This was considered to be due to the possibility that additional Se may not have increased the production of a high-priority protein such as SeP because adaptation was already ensuring that Se was going to critical pools.[13,39]

### 9.4.5 Stable-isotope Studies with High Selenium Intakes

In contrast to Finley et al., Janghorbani et al. studied the excretion of Se in men who had consumed high levels of Se during their entire lives.[11,13] Ten adult males from Jianshi County in the People's Republic of China, who had estimated daily intakes of 197 to 1230 μg Se/day, had baseline urine samples taken and then were moved to the Keshan area for 70 days. During the 70-day period, their daily Se intakes were between 30 and 45 μg/day. On day 1 of the protocol, each subject was infused by intravenous administration with 105 μg of $^{74}SeO_3^{2-}$. Twenty-four hour urine collections were made for the first seven days and thereafter at days 22, 43, and 62. An identical infusion was repeated at day 64, with the exception that the total amount of Se was 113 μg. Again, 24-hour urine collections were made for the next 7 days. Total Se and trimethylselenonium (TMeSe) were determined in all volunteers, but stable isotopes of Se as total Se and TMeSe were measured in only five men.

Urinary Se and TMeSe excretion did not differ during the first seven days after the life-long high exposure, but after the infusion at day 64, both Se and TMeSe excretion declined during the following seven days. The authors postulated that this suggested that enough Se had been depleted during the low Se intake period to affect the urinary excretion of Se and TMeSe. However, even in subjects with life-long high Se intakes, TMeSe constituted a quantitatively small fraction of urine Se; it may be that a combination of TMeSe excretion in urine combined with dimethylselenide in breath (see Figure 9.1) provides the most accurate measure of whole-body $Se^{2-}$ flux. This may be useful as a measure of the ability of whole-body Se to maintain tissue fluxes of biologically active Se for chemopreventitive purposes but further work is needed to confirm this approach.[11] The authors also noted that the specific activities (corresponding to the ratio $^{74}Se$ enriched:$^{77}Se$ unenriched) of urine Se and its TMeSe component were nearly identical, indicating a very close precursor-product relationship for these two components. This may mean that most urine Se is related to the methylation pathway of tissue $Se^{2-}$ and, equally important, indicate identical metabolism for infused $^{74}SeO_3^{2-}$ and catabolized endogenous Se. If this were to be the case, then urinary excretion of infused labelled $SeO_3^{2-}$ may also provide a means for monitoring whole body flux of endogenous $Se^{2-}$ as endogenously produced TMeSe or urine Se.[11]

When 200 μg of $^{76}SeO_3^{2-}$ was given as an oral dose with test meals containing Se as either intrinsically labelled cod ($^{82}Se$), wheat ($^{77}Se$, consumed as porridge), yeast ($^{77}Se$), or garlic ($^{82}Se$), the absorption of the selenite was found to be significantly different from the food with which it was consumed, except for yeast.[88] Selenite absorption was reported to be decreased when fed with food in an earlier study.[79] Thus the use of $SeO_3^{2-}$ as a reference dose is not recommended. Plasma appearance of the Se tracer from the cod was much slower (peaking at around 24 hours after the meal) than for plant foods (peaking at around 7 to 8 hours for wheat and garlic but later for yeast).[88] For 12 volunteers, mean apparent Se absorption (%, ± standard deviation) from garlic was 80.7 ± 11.5 and for wheat was 82.7 ± 2.65, while for cod it was 60.8 ± 5.66 and for yeast around 42.7 ± 6.61. The small interindividual variation suggests consistent absorption characteristics from each food with no homeostatic control over absorption.[88]

## 9.5 Conclusion

During the last decade, the use of stable isotopes in the study of Se metabolism and in the search for appropriate markers of status has increased steadily as the means to measure them have improved. The work has been underpinned by the vast amount of knowledge gained from animal and cell work as well as from radioisotope studies with human volunteers. The functions of the many selenoproteins are slowly being unravelled and this information, combined with more sophisticated kinetic models, should enable the Se status of humans to be better ascertained. The role of stable isotopes should be one that is carried out in conjunction with other measures of Se function wherever practicable. It is not likely that one approach will suffice, given the complexity of Se's metabolism and role in human physiology. Its behavior is related to that of iodine and vitamin E and it seems reasonable to postulate that other nutrients may also be linked with Se. The effect of Se deficiency or poor nutrient status appears to influence the way in which viruses may affect us and our propensity towards developing cancer.

However, despite the vast amount of literature concerning Se at the end of the twentieth century, "little accurate information is available on the relationship between long term Se intake in different chemical forms, the resultant body status of Se, and the potential indices that could be used to monitor Se status."[11] We still do not understand why, for example, in a country like

New Zealand, that has areas of Se intakes that are considered too low by North American and European standards, the inhabitants do not exhibit overt signs of Se depletion or higher cancer and disease rates.[12] The use of stable isotopes should enable the design of experiments to help understand the way in which humans adapt to the extremes of Se intakes found across the world. In addition to providing access to measurements which enable accurate estimates of body-pool compartments, Se isotopes also permit labelling of specific forms of the element. This should eventually permit a better understanding of the role different chemical species play in the complex biochemistry of Se.

## References


1. Schwartz, K. and Foltz, C.M., Selenium as an integral part of factor 3 against necrotic liver degeneration, *J. Amer. Chem. Soc.*, 79, 3292, 1957.
2. Rotruck, J.T. et al., Selenium: biochemical role as a component of glutathione peroxidase, *Science*, 179, 588, 1973.
3. Reilly, C., *Selenium in Food and Health*, Blackie Academic & Professional, London, 1996, chap.1.
4. Ysart, G. et al., Dietary exposure estimates of 30 elements from the U.K. Total Diet Study, *Fd. Add. Contam.*, 16, 391, 1999.
5. Ysart, G. et al., 1997 U.K. Total Diet Study — dietary exposures to aluminium, arsenic, cadmium, chromium, copper, lead, mercury, nickel, selenium, tin and zinc, *Fd. Add. Contam.*, 17, 775, 2000.
6. Verlinden, M. et al., The selenium status of Belgian population groups. 1. Healthy adults, *Biol. Trace Elem. Res.*, 5, 91, 1983.
7. Kumpulainen, J. and Tahvonen, R., Report on the activities of the sub-network on trace element status in food, in *Report of the 1989 Consultation of the European Cooperative Research Network on Trace Elements*, Lausanne, Switzerland, 5–8 Sept. 1989, FAO, Rome, 1989.
8. Parr, R.M. et al., Dietary intakes of trace elements and related nutrients in eleven countries: preliminary results from an IAEA coordinated research programme, in *Proceedings of the 7th International Symposium on Trace Elements in Man and Animals — TEMA7*, Dubrovnik, Yugoslavia, 20–25, May 1990, Momcilovic, B., Ed., IMI, Zagreb, 1990, 13/3.
9. Yang, G.Q. et al., Endemic selenium intoxication of humans in China, *Am. J. Clin. Nutr.*, 37, 872, 1983.
10. Yang, G.Q. et al., Studies of safe maximal daily dietary Se-intake in a seleniferous area in China. Part II: relation between Se-intake and the manifestation of clinical signs and certain biochemical alterations in blood and urine, *J. Trace Elem. Electrolytes Health Dis.*, 3, 123, 1989.
11. Janghorbani, M. et al., Quantitative significance of measuring trimethylselonium in urine for assessing chronically high intakes of selenium in human subjects, *Br. J. Nutr.*, 82, 291, 1999.
12. Robinson, M.F., 1988 McCollum Award lecture. The New Zealand selenium experience, *Am. J. Clin. Nutr.*, 48, 521, 1988.

13. Finley, J.W. et al., Selenium supplementation affects the retention of stable isotopes of selenium in human subjects consuming diets low in selenium, *Br. J. Nutr.*, 82, 357, 1999.
14. Golubkina, N.A. and Alfthan, G.V., The human selenium status in 27 regions of Russia, *J. Trace Elements Med. Biol.*, 13, 15, 1999.
15. Whanger, P.D., Metabolism of selenium in humans, *J. Trace Elem. Exp. Med.*, 11, 227, 1998.
16. Deagen, J.T. et al., Effects of dietary selenite, selenocystine, and selenomethionine on selenocysteine lyase and glutathionine peroxidase activities and on selenium levels in rat tissues, *J. Nutr.*, 117, 91, 1987.
17. Vendeland, S.C., Butler, J.A., and Whanger P.D., Intestinal absorption of selenite, selenate, and selenomethionine in the rat, *J. Nutr. Biochem.*, 3, 359, 1992.
18. Fairweather-Tait, S.J., Bioavailability of selenium, *Eur. J. Clin. Nutr.*, 51, S20, 1997.
19. Moesgaard, S. and Morrill, R., The need for speciation to realise the potential of selenium in disease prevention, in *Trace Element Speciation for Environment, Food and Health,* Cornelis R., Crews, H., Donard, O., Ebdon, L., and Pitts, L., Eds., Royal Society of Chemistry, Cambridge, in press.
20. Jackson, M. J., The assessment of the bioavailability of micronutrients: introduction, *Eur. J. Clin. Nutr.*, 51, S1, 1997.
21. Crews, H.M., Speciation of trace elements in foods, with special reference to cadmium and selenium: is it necessary?, *Spectrochimica Acta Part B*, 53, 213, 1998.
22. Suzuki, K.T., Itoh, M., and Ohmichi, M., Detection of selenium-containing biological constituents by high-performance liquid chromatography-plasma source mass spectrometry, *J. Chromatogr. B*, 666, 13, 1995.
23. Reilly, C., Biological roles of selenium, in *Selenium in Food and Health*, Reilly, C., Ed., Blackie Academic & Professional, London, 1996, chap.2.
24. Behne, D. et al., Effects of chemical form and dosage on the incorporation of selenium into tissue proteins in rats, *J. Nutr.*, 121, 806, 1991.
25. Wilhelmsen, E.C., Hawkes, W.C., and Tappel, A.L., Substitution of selenocysteine for cysteine in a reticulocyte lysate protein synthesis system, *Biol. Trace Elem. Res.*, 7, 141, 1985.
26. Patching, S.G. and Gardiner, P.H.E., Recent developments in selenium metabolism and chemical speciation: a review, *J. Trace Elements Med. Biol.*, 13, 193, 1999.
27. Arthur, J.R. and Beckett, G.J., New metabolic roles for selenium, *Proc. Nutr. Soc.*, 53, 615, 1994.
28. World Health Organization, *Selenium*, Environmental Health Criteria 58, WHO, Geneva, 1987, chap. 8.
29. Parízek, J., Health effects of dietary selenium, *Fd. Chem. Toxic.*, 28, 763, 1990.
30. Beck, M.A., Selenium and host defence towards viruses, *Proc. Nutr. Soc.*, 58, 707, 1999.
31. Turner, R.J. and Finch, J.M., Selenium and the immune response, *Proc. Nutr. Soc.*, 50, 275, 1991.
32. Diplock, A.T., Indexes of selenium status in human populations, *Am. J. Clin. Nutr. Suppl.*, 57, 245S, 1993.
33. Ip, C. and Lisk, D.J., Characterisation of tissue selenium profiles and anticarcinogenic responses in rats fed natural sources of selenium-rich products, *Carcinogenesis,* 15, 573, 1994.
34. Clark, L.C. et al., Effect of selenium supplementation for cancer prevention in patients with carcinoma of the skin, *J. Am. Med. Assoc.*, 276, 1957, 1996.

35. Department of Health, Selenium, in *Dietary Reference Values for Food Energy and Nutrients for the United Kingdom, Report on Health and Social Subjects No 41*, HMSO, London, 1991, 174.
36. National Research Council, *RDA — Recommended Dietary Allowances*, NRC National Academy Press, Washington D.C., 1989, 220.
37. Reilly, C., Selenium in health and disease: a review, *Aust. J. Nutr. Diet.*, 50, 136, 1993.
38. Arthur, J.R., Functional indicators of iodine and selenium status, *Proc. Nutr. Soc.*, 58, 507, 1999.
39. Sunde, R.A., Intracellular glutathione peroxidases — structure, regulation and function, in *Selenium in Biology and Human Health*, Burk, R.F., Ed., Springer-Verlag, New York, 1994, chap. 3.
40. Mills, G.C., Hemoglobin catabolism. I. Glutathione peroxidase, an erythrocyte enzyme which protects hemoglobin from oxidative breakdown, *J. Biol. Chem.*, 229, 189, 1957.
41. Behne, D. et al., Studies in the distribution and characteristics of new mammalian selenium-containing proteins, *Analyst*, 120, 823, 1995.
42. Cohen, H.J. and Avissar, N., Extracellular glutathione peroxidase: a distinct selenoprotein, in *Selenium in Biology and Human Health*, Burk, R.F., Ed., Springer-Verlag, New York, 1994, chap. 4.
43. Yang, G. et al., Human selenium requirements in China, in *Selenium in Biology and Medicine, Part B*, Combs, G.F., Spallholtz, J.E., Levander, O.A., and Oldfield, J.E., Eds., AVI Publishing Co., Westport, CT, 1987, 589.
44. Arthur, J.R. et al., Regulation of selenoprotein gene expression and thyroid hormone metabolism, *Biochem. Soc. Trans.*, 24, 384, 1996.
45. Gladyshev, V.N., Jeang, K.Y., and Stadtman, T.C., Selenocysteine, identified as the penultimate C-terminal residue in human T-cell thioredoxin reductase, corresponds to TGA in the human placental gene, *Proc. Nat. Acad. Sci. U.S.A.*, 93, 6146, 1996.
46. Howie, A.F. et al., Identification of a 57-kilodalton selenoprotein in human thyrocytes as thioredoxin reductase and evidence that its expression is regulated through the calcium-phosphoinositol signalling pathway, *J. Clin. Endoc. Metab.*, 83, 2052, 1998.
47. Hill, K.E. and Burk, R.F., Selenoprotein P — an extracellular protein containing multiple selenocysteines, in *Selenium in Biology and Human Health*, Burk, R.F., Ed., Springer-Verlag, New York, 1994, chap. 6.
48. Burk, R.F. et al., Response of rat selenoprotein P to selenium administration and fate of its selenium, *Am. J. Physiol.*, 261, E26, 1991.
49. Mellon, F.A. and Sandström, B., *Stable Isotopes in Human Nutrition, Inorganic Nutrient Metabolism*, Academic Press Ltd., London, 1996, 123.
50. Nève, J. and Carpentier, Y.A., Laboratory methods for the assessment of selenium status, in *Nutrition in Clinical Practice, Proc. 10th Congr. ESPEN, Leipzig 1988*, Hartig, Dietze, Weiner, and Furst, Eds., Karger, Basel, 1989, 273.
51. Crews. H.M. et al., The analyst's viewpoint with special reference to selenium, *Nutr. Fd. Sci.*, 6, 221, 1997.
52. Paglia, D.E. and Valentine, W.N., Studies on the quantitative and qualitative characterisation of erythrocyte glutathione peroxidase. *J. Lab. Clin. Med.*, 70, 158, 1967.
53. Wendel, A., Glutathione peroxidase. *Meth. Enzymol.*, 77, 325, 1981.
54. Faraji, B., Kang, H.K,, and Valentine, J.L., Methods compared for determining glutathione-peroxidase activity in blood, *Clin. Chem.*, 33, 539, 1987.

55. Whanger, P.D. et al., Blood selenium and glutathione-peroxidase activity of populations in New Zealand, Oregon, and South-Dakota, *FASEB J.*, 2, 2996, 1988.
56. Huang, W. and Åkesson, B., Radioimmunoassay of glutathione peroxidase in human serum, *Clin. Chim. Acta.*, 219, 139, 1993.
57. Huang, W. et al., Selenoprotein P and glutathione peroxidase (EC1.11.1.9) in plasma as indices of selenium status in relation to the intake of fish, *Brit. J. Nutr.*, 75, 455, 1995.
58. Marchaluk, E. et al., Variation in selenoprotein P concentration in serum from different European regions, *Eur. J. Clin. Nutr.*, 49, 42, 1995.
59. Nève, J., Methods in determination of selenium status, *J. Trace Elem. Electrol. Health Dis.*, 5, 1, 1991.
60. Alfthan, G. et al., Selenium metabolism and platelet glutathione-peroxidase activity in healthy Finnish men — effects of selenium yeast, selenite, and selenate, *Am. J. Clin. Nutr.*, 53, 120, 1991.
61. Fox, T. et al., Changes in indices of selenium status in men on low, medium and high intakes, in *Trace Elements in Man and Animals — 10 Proceedings, TEMA-10 Evian May 1999*, Favier, A., Anderson, R.A., and Roussel A.M., Eds., Plenum, New York, 2000, in press.
62. Alt, F. and Messerschmidt, J., Selenium in *Quantitative Trace Analysis of Biological Materials*, McKenzie, H.A. and Smythe, L.E., Eds., Elsevier Science Publishers B.V., Amsterdam, The Netherlands; 1988, 487.
63. Thomassen, Y. and Aaseth, J., Human tissues, in *Occurrence and Distribution of Selenium*, Ihnat, M., Ed., CRC Press, Boca Raton, 1989, 169–212.
64. Veillon, C., GC/MS measurement of stable isotopes of selenium for use in metabolic tracer studies, in *Stable Isotopes in Nutrition*, Turnlund, J.R. and Johnson, P.E., Eds, American Chemical Society, USA, 1984, chap. 7.
65. Crews, H.M. et al., Mass spectrometric methods for studying nutrient mineral and trace element absorption and metabolism in humans using stable isotopes, *Analyst*, 119, 2491, 1994.
66. Ducros, V., Electron ionization mass spectrometry (EIMS) and gas chromatography-mass spectrometry (GC-MS), in *Stable Isotopes in Human Nutrition, Inorganic Nutrient Metabolism*, Mellon, F.A. and Sandström, B., Eds., Academic Press Ltd., London, 1996, chap. 9.
67. Janghorbani, M. and Ting, B.T.G., Stable isotope tracer applications of ICP-MS, in *Applications of Inductively Coupled Plasma-mass Spectrometry*, Date, A.R. and Gray, A.L., Eds., Blackie, Glasgow and London, UK, 1989, chap. 5.
68. Crews, H.M., Luten, J.B., and McGaw, B.A., Inductively coupled plasma-mass spectrometry, in *Stable Isotopes in Human Nutrition, Inorganic Nutrient Metabolism*, Mellon, F.A. and Sandström, B., Eds., Academic Press Ltd., London, 1996, chap. 12.
69. Martin, R.F., Janghorbani, M., and Young, V.R., Kinetics of a single administration of $^{74}$Se-selenite by oral and intravenous routes in adult humans, *J. Parent. Ent. Nutr.*, 12, 351, 1988.
70. Sandström, B. et al., Methods for studying mineral and trace element absorption in humans using stable isotopes, *Nutr. Res. J.*, 6, 71, 1993.
71. Yoshinaga, J. et al., Isotope dilution analysis of selenium in biological materials by nitrogen microwave-induced plasma mass spectrometry, *Anal. Chem.*, 67, 1568, 1995.
72. Ting, B.T.G., Mooers, C.S., and Janghorbani, M., Isotopic determination of selenium in biological materials with inductively coupled plasma-mass spectrometry, *Analyst*, 114, 667, 1989.

73. Buckley, W.T. et al. Determination of multiple selenium stable isotope tracers by inductively coupled plasma-mass spectrometry, *Biol. Mass Spectrom.*, 21, 473, 1992.
74. Crews, H.M. et al., Investigation of selenium speciation in *in vitro* gastro-intestinal extracts of cooked cod by high-performance liquid chromatography-inductively coupled plasma-mass spectrometry and electrospray mass spectrometry, *J. Anal. At. Spectrom.*, 1996, 11, 1177, 1996.
75. Aggarwal, S.K., Kinter, M., and Herold, D.A., Determination of selenium in urine by isotope dilution gas chromatography-mass spectrometry using 4-nitro-*o*-phenylenediamine, 3, 5 dibromo-*o*-phenylenediamine, and 4-trufluoromethyl-*o*-phenylenediamine as derivatising agents, *Anal. Biochem.*, 202, 367, 1992.
76. Smink, N. et al., Analysis of selenium stable isotope ratios in urine and faeces using hydride generation-inductively coupled plasma-mass spectrometry: Results of an interlaboratory comparison (EC contract FAIR-CT-95-0771), *J. Anal. At. Spectrom.*, 2000, to be submitted.
77. van Dokkum, W. et al., Study Techniques, in *Stable Isotopes in Human Nutrition, Inorganic Nutrient Metabolism*, Mellon, F.A. and Sandström, B., Eds., Academic Press Ltd., London, 1996, chap. 4.
78. Wilson, P.D.G. and Dainty, J.R., Modelling in nutrition: an introduction, *Proc. Nutr. Soc.*, 58, 133, 1999.
79. Christensen, M.J. et al., Simultaneous determination of absorption of selenium from poultry meat and selenite in young men: application of a triple stable isotope method, *Br. J. Nutr.*, 50, 43, 1983.
80. Sirichakwal, P.P., Young, V.R., and Janghorbani, M., Absorption and retention of selenium from intrinsically labelled egg and selenite as determined by stable isotope studies, *Am. J. Clin. Nutr.*, 41, 264, 1985.
81. Janghorbani, M., Kasper, L.J., and Young, V.R., Dynamics of selinite metabolism in young men: studies with the stable isotope tracer method, *Am. J. Clin. Nutr.*, 40, 208, 1984.
82. Janghorbani, M. et al., The selenite exchangeable pool in humans: a new concept for the assessment of selenium status, *Am. J. Clin. Nutr.*, 51, 670, 1990.
83. Rennie, M.J., An introduction to the use of tracers in nutrition and metabolism, *Proc. Nutr. Soc.*, 58, 935, 1999.
84. Westerterp, K.R., Body composition, water turnover and energy turnover assessment with labelled water, *Proc. Nutr. Soc.*, 58, 945, 1999.
85. Coggan, A.R., Use of stable isotopes to study carbohydrate and fat metabolism at the whole-body level, *Proc. Nutr. Soc.*, 58, 953, 1999.
86. Swanson, C.A. et al., Quantitative and qualitative aspects of selenium utilisation in pregnant and non pregnant women: an application of stable isotope methodology, *Am. J. Clin. Nutr.*, 38, 169, 1983.
87. Finley, J., The retention and distribution by healthy young men of stable isotopes of selenium consumed as selenite, selenate or hydroponically-grown broccoli are dependent on the isotopic form, *J. Nutr.*, 129, 854, 1999.
88. Atherton, C. et al., Absorption of selenium from biosynthetically labelled foods in humans, in *Trace Elements in Man and Animals — 10 Proceedings, TEMA-10 Evian May 1999*, Favier, A., Anderson, R.A., and Roussel A.M., Eds., Plenum, New York, 2000, in press.

# 10

# *Use of Isotopes for Studies with Manganese, Chromium, and Molybdenum*

**John W. Finley**

**CONTENTS**


10.1 Manganese ....... 152
  10.1.1 Introduction ....... 152
  10.1.2 Manganese Biochemistry ....... 152
  10.1.3 Radioactive Isotopes of Manganese and Studies of Manganese Essentiality ....... 153
    10.1.3.1 Studies with Laboratory Animals and Cultured Cells .. 153
    10.1.3.2 Distribution and Retention of Radioactive Manganese in Humans ....... 154
    10.1.3.3 Radioactive Methods of Determining Apparent Manganese Absorption in Humans ....... 155
    10.1.3.4 Radioactive Methods for Determining True Manganese Absorption ....... 156
    10.1.3.5 The Use of Radioisotopes to Study Manganese/Iron Interactions ....... 158
10.2 Chromium ....... 159
  10.2.1 Introduction ....... 159
  10.2.2 Chemistry and Biochemistry ....... 160
  10.2.3 Radioactive Chromium in Human Studies ....... 160
    10.2.3.1 Nutritional Studies with $^{51}Cr$ ....... 160
    10.2.3.2 Stable Isotopes of Chromium in Human Studies ....... 161
10.3 Molybdenum ....... 161
  10.3.1 Chemistry and Biochemistry ....... 161
  10.3.2 Radioactive Isotopes of Molybdenum in Human Studies ....... 161
  10.3.3 Stable Isotopes of Molybdenum in Human Studies ....... 162
10.4 Summary ....... 162
References ....... 163


0-8493-8730-2/00/$0.00+$.50

## 10.1 Manganese

### 10.1.1 Introduction

Manganese (Mn) has a single stable isotope; therefore, experimental protocols requiring the use of a tracer must use a radioisotope. There are multiple radioactive isotopes of Mn, but only three have been used for biological research. The primary isotope used in biological studies, $^{54}$Mn, is a gamma-emitter with a single unique energy of 835 KEV that decays by electron capture with a half-life of 312 days. Radioactive $^{52}$Mn and $^{56}$Mn also have been used in research; however, relatively short half-lives (5.6 days for $^{52}$Mn and 2.6 hours for $^{56}$Mn) and poor availability ($^{52}$Mn must be custom-produced by a cyclotron reaction and the short half-life of $^{56}$Mn requires it to be produced at the site of the experiment) have made them unacceptable for most routine biological applications.

### 10.1.2 Manganese Biochemistry

Manganese is studied in living systems because it is essential for life and because excessive Mn is toxic. Manganese is essential because it is required by several mammalian enzymes. Manganese deficiency is well documented and causes severe health problems in domestic animals, but there are only a few reports of human Mn deficiency. A recommended dietary allowance (RDA) for Mn has not been determined and is an area that needs continued research.[1]

In contrast to the paucity of reports of human Mn deficiency, there are many reports of acute toxicity in human miners that inhale Mn-laden dust.[2] Recent reports suggest that Mn toxicity also may be a problem in people with hepatic insufficiency who consume normal amounts of Mn.[3] This, too, is an area that needs further research attention. Tracer forms of Mn are essential for many studies of both Mn deficiency and excess.

Animals consume milligram amounts of dietary Mn on a daily basis (2 to 10 mg/day represents a normal range of intake), and a very small percentage (<10%) is apparently absorbed along the length of the intestinal tract.[4] There is controversy as to whether absorption is active or by diffusion.[5] Several nutrients and food components seem to affect the extent of Mn absorption.[6] Absorbed Mn enters the portal system bound to a ligand that may be albumin or transferrin, and is rapidly taken up by the liver and secreted into the bile against a strong concentration gradient.[7–9] Maximum biliary excretion of intravenously injected manganese in rats occurs 15 to 60 minutes after injection.[10] Manganese escaping first-pass liver clearance enters the systemic circulation bound primarily to transferrin as $Mn^{3+}$; in the blood, manganese is distributed to tissues, where it functions as a component or activator of numerous enzymes.[8,11]

### 10.1.3 Radioactive Isotopes of Manganese and Studies of Manganese Essentiality

#### *10.1.3.1 Studies with Laboratory Animals and Cultured Cells*

Radioactive Mn has been used to study Mn absorption, tissue distribution, hepatic metabolism, and basic biochemistry.[5,12–18] Finley and Monroe[19] studied Mn uptake and transcellular movement in a cell culture model of the enterocyte. Mathematical analysis of kinetic data of $^{54}$Mn transcellular movement was used to estimate basic parameters of absorption including km, Jmax, a diffusion constant and the effects of various inhibitors. The rate of $^{54}$Mn transcellular movement was greater in a basolateral to apical direction than in an apical to basolateral direction, a finding suggesting that the enterocyte may actively excrete Mn into the gut lumen. Hepatic metabolism of $^{54}$Mn has been studied in cultured human hepato-carcinoma cells and in isolated rat hepatocytes.[9,15] Hepatocytes rapidly incorporated and excreted Mn by a mechanism that may involve Ca channels.

Radioactive $^{54}$Mn has been used in many studies with laboratory animals; this portion of the review will focus on whole-animal techniques with application to human studies. Strause et al. gavaged mice with 0.11 mBq of $^{54}MnCl_2$ complexed to nitrilotriacetate and reported that less than 3% of the administered dose remained after 10 days.[20] Weigand et al. injected rats intramuscularly with $^{54}$Mn as $MnCl_2$.[21] Based on tissue retention of $^{54}$Mn, they concluded that true Mn absorption was substantially greater than apparent absorption and the intestine was the major organ involved in Mn homeostasis.

Detection of whole-body counts (WBC) of $^{54}$Mn in laboratory animals is a powerful technique for estimation of Mn absorption, retention, and excretion. Lee and Johnson used a custom-built, whole-body counter equipped with two thallium-activated NaI detectors and a ND62 multichannel analyzer (Nuclear Data Instrumentation, Schaumberg, IL*) with 2048 channels.[22] The counter was calibrated with $^{22}$Na, and a $^{54}$Mn standard was counted daily to adjust for fluctuations of counting efficiency. All data were adjusted for background and radioactive decay. *In vivo* absorption was calculated as the y-intercept of the linear portion of a semi-logarithmic plot of (decay-corrected) whole-body counts vs time after isotope administration. Biological half-life was calculated as –ln2/slope of the same portion of the curve. These techniques were used to study Mn homeostasis in rats fed diets high or low in Mn and administered 0.73 mBq of $^{54}$Mn by gavage, test meal, intramuscular injection or intra-peritoneal injection.[22] Less $^{54}$Mn was absorbed and biological half-life was shorter in rats fed the low, as compared to the high, Mn diet. Test meals labelled with $^{54}$Mn and mathematical analysis of WBC kinetic data also have been used to study the influence of dietary components on Mn bioavailability to rats, the effects of sex and age of mice on Mn absorption and biological half-life, the effect of biliary ligation on retention of $^{54}$Mn in rats, the

---

* Mention of a trademark or proprietary product does not constitute a guarantee or warranty of the product by the United States Department of Agriculture and does not imply its approval to the exclusion of other products that may also be suitable.

influence of a genetic defect of Fe metabolism on $^{54}$Mn and $^{59}$Fe absorption and retention, and the effect of fat type on Mn absorption (unreported data).[23–26]

Diez-Ewald et al. studied the dietary interaction of Mn and Fe by perfusing $^{54}$Mn through isolated gut loops of rats previously gavaged with a $^{59}$Fe (ferrous)-citrate complex.[27] Thomson et al. perfused the intestines of Fe-deficient rats with 1.11 kBq of carrier-free $^{54}MnCl_2$ with ascorbate present in a 1:2 manganese:ascorbate molar ratio to prevent Mn from precipitating from the solution.[28] Both studies concluded that Mn absorption is depressed when Fe is added to the gut, and that Mn absorption is improved by Fe deficiency.

Simultaneous use of $^{59}$Fe and $^{54}$Mn requires the energy of each nuclide to be corrected for "spill-over" energy from the other nuclide, i.e., energy from one nuclide that is detected within the window of detection of the other nuclide. Roughead et al. used WBC to follow the retention of $^{59}$Fe and $^{54}$Mn in mice.[26] Gamma energies of $^{54}$Mn and $^{59}$Fe were measured in separate detection windows (700 to 890 and 980 to 1390 keV, respectively) of a small-animal, whole-body counter. Corrections were made by measuring the energies in three sets of standard solutions containing 1) $^{54}$Mn, 2) $^{59}$Fe, or 3) $^{54}$Mn and $^{59}$Fe. The contribution of $^{59}$Fe to the $^{54}$Mn window was calculated by comparison of these standard curves and $^{59}$Fe contributed approximately 21.5% to $^{54}$Mn energies. This correction was valid only over the range where the contribution of one isotope to the other was linear. The contribution of $^{59}$Fe to $^{54}$Mn in a well-type gamma counter was calculated to be 30.1%, which illustrates that the correction factor is unique for each instrument.

#### *10.1.3.2 Distribution and Retention of Radioactive Manganese in Humans*

The only radioactive isotope of Mn widely available for human studies is $^{54}$Mn, although past investigators have also used $^{52}$Mn and $^{56}$Mn.[29,30] The relatively long half-life of $^{54}$Mn, its relatively strong gamma energy, and the relatively great affinity of Mn to some tissues restricts the amount of $^{54}$Mn that can be used in human studies. Because of the small amounts of $^{54}$Mn that can be administered to humans, and because most of a dose of $^{54}$Mn is quickly eliminated in the feces, there is very little detectable $^{54}$Mn in most organs and tissues. Consequently, unless large amounts of tissues and organs can be collected, human studies with $^{54}$Mn are restricted to facilities with whole-body counters capable of detecting very low amounts of whole-body $^{54}$Mn radiation.

Sheppard et al. intravenously administered $^{52}$Mn (custom produced in a research reactor) to terminally ill patients receiving internal radiation therapy.[31] Autopsy tissues were digested and counted by a Geiger-Müller counter, and the liver, kidney, and pancreas were reported to contain the most $^{52}$Mn. Borg and Cotzias studied the distribution and blood transport of Mn in humans injected intravenously with $^{56}$Mn.[30] This research was possible because it was conducted in the Medical Department of Brookhaven National Laboratory, which allowed on-site production of the isotope. Aliquots of 0.56 to 0.74 mBq were prepared, and approximately half of the radioactivity remained at the time of injection. Sampled tissues were counted in a well-type gamma counter and whole-body radiation was estimated by a whole-body counter that consisted of several collimated detectors held over the body surface; radiation was detected throughout 72 hours. The kinetics of $^{56}$Mn retention in several organs and tissues including blood, liver, and thigh were described, and the rapid clearance of Mn from the blood and into the liver was hypothesized to be the result of rapid mitochondrial accumulation.

Cotzias and coworkers studied the effects of Mn poisoning on $^{54}$Mn retention and excretion in miners.[32] Intravenously administered $^{54}$Mn was cleared from the blood with a half-life of 1.3 to 2.2 minutes, and the rate of clearance was enhanced in miners with Mn poisoning. Mena et al. administered by inhalation 3.7 mBq of $^{54}Mn_2O_3$ or $^{54}MnCl_2$ to normal or Mn-exposed miners.[33] They reported rapid transfer (within several days) of the $^{54}$Mn from the lungs to the G.I. tract and, within 4 days, 60% of the inhaled radiation was recovered in the feces. Mena et al. also showed evidence for the inhibition by Fe of Mn absorption by orally administering an aqueous solution of $^{59}$Fe citrate (approximately .37 mBq) and $^{54}MnCl_2$ (approximately 3.7 mBq) to subjects followed by simultaneous detection of WBC of $^{59}$Fe and $^{54}$Mn.[33] The Compton contribution of $^{59}$Fe to the 0.84 MEV peak of $^{54}$Mn was calculated and subtracted before analysis of results.

#### *10.1.3.3 Radioactive Methods for Determining Apparent Manganese Absorption in Humans*

Mena et al. estimated Mn absorption from the whole-body retention of $^{54}$Mn 72 hours after consumption of a radioactive test meal.[33] Absorption was estimated to be less than 10%; however, the authors noted inherent problems with predicting absorption from a steeply declining portion of the whole-body retention curve. They also addressed difficulties in determining true absorption for a nutrient that undergoes entero-hepatic recirculation. Mahoney and Small described "fast" and "slow" components of whole-body retention curves (biological half-lives of 4 days and 39 days, respectively) in subjects that orally consumed 0.09 mBq $^{54}$Mn.[34] Whole-body $^{54}$Mn was measured for 90 days in an iron-surrounded WBC facility with two thallium-activated NaI crystal detectors.

Davidsson et al. used a room with iron walls, two NaI crystals, and four plastic scintillator block detectors that lay above and below the subject to measure whole-body $^{54}$Mn.[35] A dose of 0.19 (oral administration) or 0.07 mBq of $^{54}$Mn (i.v. administration) was given simultaneously with 1.5 mBq of $^{51}$Cr that was used as a non-absorbable marker. They reported large inter-individual variation in the rate of Mn excretion and concluded that the best way to measure factors affecting Mn retention was to give repeated doses of $^{54}$Mn to the same individual. They also reported that less variability was associated with the slow turnover component than the fast component, and the slow turnover component could be fit with a single exponential model. Absorption was estimated by several methods including by comparison of $^{54}$Mn in a single fecal sample to $^{51}$Cr that was given as a non-absorbable marker. However, results of this method were variable and resulted in a range of –19 to 25%. Absorption was also calculated by extrapolation of the linear portion of the WBC retention curve back to y-axis; a method that was more precise and gave values that ranged from 0.8 to 16%, with most measures falling between 2 and 7%.

Davidsson et al. fed young women the livers from chickens previously gavaged with $^{54}Mn$, or unlabelled chicken livers with $^{52}Mn$ pipetted onto the surface.[36] Intrinsically labelled meals contained 0.1 MBq of $^{54}Mn$ and extrinsically labelled meals contained 0.03 to 0.4 MBq of $^{52}Mn$. $^{52}Mn$ was produced by cyclotron irradiation of Cr foil, followed by chemical separation. Mathematical corrections were made to account for spillover in the peaks between the two nuclides, and radiation from both nuclides was detected during the next 30 days. Intrinsic and extrinsic labelling resulted in virtually identical patterns of whole-body retention of radiation.

Johnson and coworkers fed healthy subjects spring wheat, confectionary sunflowers, lettuce, and spinach intrinsically labelled by stem injection or extrinsically labelled by adding $^{54}Mn$ solution to these foods.[37] Test meals contained 0.037 mBq $^{54}Mn$, and gamma radiation was detected in a steel-enclosed, air-filtered chamber with an array of 32 NaI detectors located above and beneath a bed. Whole-body $^{54}Mn$ was detected for eight weeks and absorption was calculated from the y-intercept of a line extrapolated from the portion of the WBC curve between days 10 to 56 (Figure 10.1a). Mean absorption was 1.7 to 5.2%, and was not different for intrinsically or extrinsically labelled foods. Johnson and Lykken fed healthy young women 0.037 mBq of $^{54}MnCl_2$ simultaneous with $^{47}CaCl_2$.[38] Whole-body retention of both nuclides was followed for 35 days. Absorption and biological half-life were calculated as before and Mn absorption was calculated to be 2.1 to 4.4%.

#### 10.1.3.4 Radioactive Methods for Determining True Manganese Absorption

Davidsson et al. suggested that problems associated with determining true absorption of Mn occur because a large amount of Mn may be absorbed and quickly excreted in the bile, while other Mn may undergo enterohepatic circulation.[35] A perfusion study in humans found an average of 27% of Mn was removed from the gut lumen, and this was increased to 67% in patients with low Fe stores.[28]

Several methods, including mathematical modelling, have been used to attempt to circumvent the problems associated with determining $^{54}Mn$ absorption. Many researchers use the linear terminal portion of $^{54}Mn$ WBC retention curves to estimate absorption and biological half-life (Figure 10.1a). They assume the curve is linear after day 10, but Finley and Johnson fed young women $^{54}Mn$ and counted whole-body radiation through 70 days.[39] Data from days 10 to 20 resulted in a biological half-life of 15.3 days and absorption of 1.35%, whereas data from days 19 to 70 resulted in a biological half-life of 48 days and absorption of 0.75%.

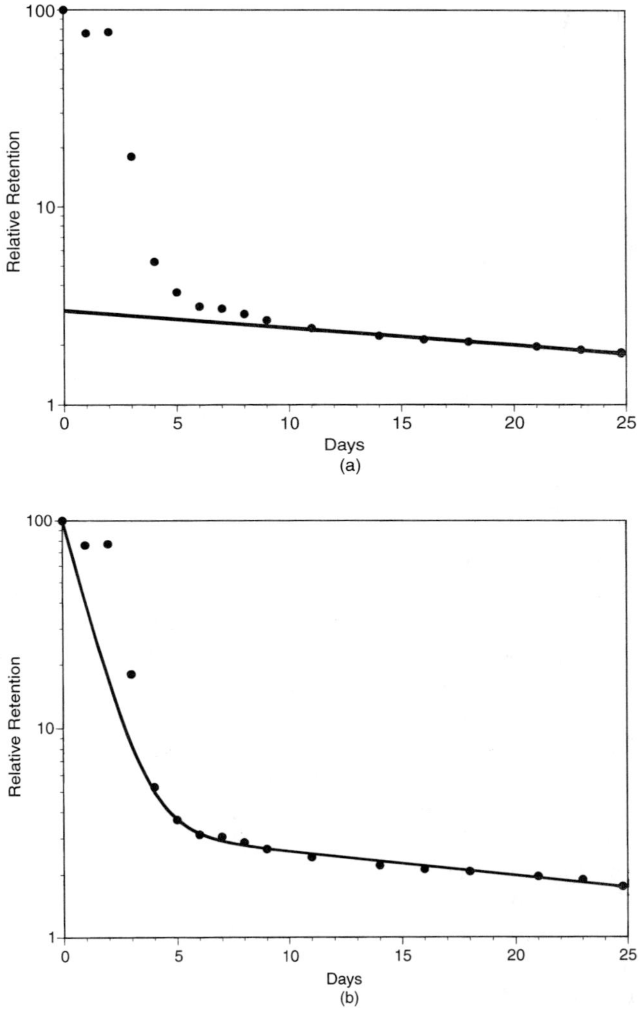


**FIGURE 10.1**
Mathematical modelling of whole-body retention of $^{54}$Mn kinetic data. Whole-body counts (WBC) are from young women that consumed a test meal containing 0.037 mBq of $^{54}$Mn. Whole-body radiation was detected in a steel-enclosed chamber containing 32 NaI detectors. Whole-body radiation is given as percent of total dose where the total dose was defined as the average of 4 whole-body counts conducted the same day as the test meal was consumed. (a) Linear modelling of the tail portion (counts after day 10) of WBC data. Absorption is predicted by extrapolating the line to the y-intercept. (b) Fit of a double-exponential model [%retn = 96.622∗exp(–0.974∗Days) + 3.378∗exp(–0.0266∗Days)] to WBC data.

*(continued)*

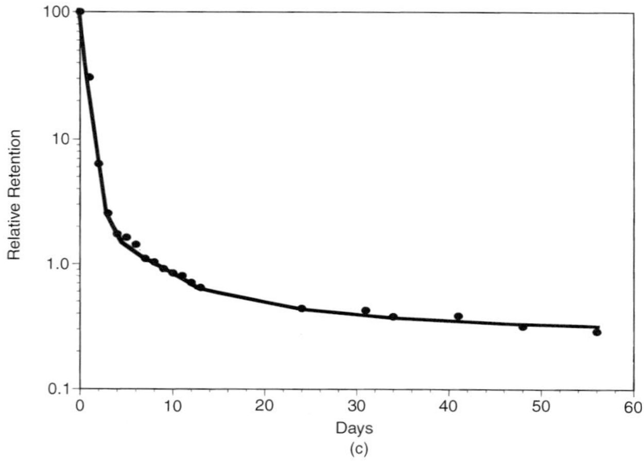


**FIGURE 10.1** ***(continued)***
(c) Fit of a six compartment model to WBC data.

Likewise, exponential models do not provide a good fit to $^{54}$Mn WBC data. Figure 10.1b shows the fit of double exponential models to whole-body retention of $^{54}$Mn by a subject fed a low-Mn diet. Such models provide a good fit to the tail portion of the curve, but greatly underestimate retention through day 4. Compartmental modelling has been used to model kinetic data of other trace elements.[40–43] Davis and coworkers used a compartmental model to describe Mn metabolism in rats and they concluded that 37% of dietary Mn was absorbed, but most was quickly lost through biliary excretion (providing more evidence for the hypothesis of absorption followed by rapid biliary excretion and enterohepatic re-circulation).[7] However, the rat may not be a good model of biliary excretion of Mn for humans. No evidence was found to support this hypothesis when $^{54}$Mn was given to surgically altered pigs that allowed for simultaneous detection of $^{54}$Mn in the portal blood supply and in the bile.[45]

Finley et al. described a preliminary six-compartment model to model the retention of $^{54}$Mn in humans (Figure 10.2).[44] The model contained multiple liver compartments with unique turnover rates, and a component representing biliary excretion of Mn. The model provided excellent fits to individual whole-body retention curves (Figure 10.1c). However, there were not sufficient pools containing measurable $^{54}$Mn to adequately validate the model.

#### *10.1.3.5 The Use of Radioisotopes to Study Manganese/Iron Interactions*

Thomson et al. studied Mn absorption in normal patients and patients with low Fe stores, and demonstrated that $^{54}$Mn disappeared from the perfused gut faster in subjects with low Fe stores.[28] Finley and Johnson found that women absorbed more $^{54}$Mn than men, but biological half-life was also shorter in women.[39] A subsequent study fed test meals containing similar

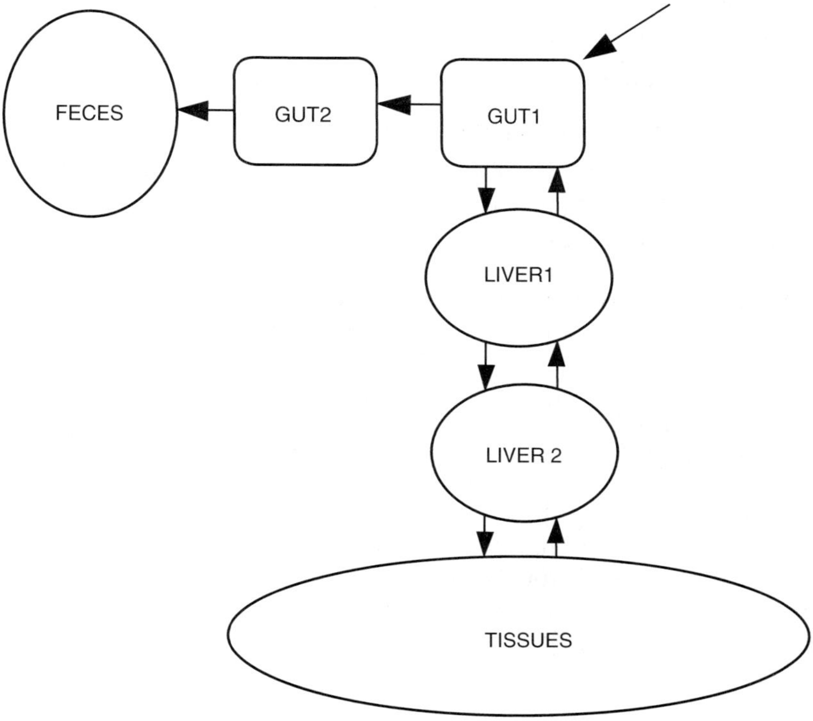


**FIGURE 10.2**
Generalized six-compartment model of manganese metabolism. Boxes represent pools, arrows represent rate transfer components.

amounts of $^{54}$Mn to healthy young women selected for low or high Fe stores.[6] Women with low Fe stores had enhanced Mn absorption but a shorter biological half-life for Mn.

## 10.2 Chromium

### 10.2.1 Introduction

Chromium (Cr) is a trace element that has been studied since 1957 because of its beneficial and essential functions in humans. Chromium was originally isolated as a "glucose tolerance factor" because of its ability to potentiate insulin.[46] Today chromium supplements are sold widely throughout North America and Europe and claims have been made that Cr redistributes metabolic energy and allows for synthesis of more muscle and less fat. Chromium has also been used in nutritional studies as a non-absorbable marker of the movement of chyme in the digestive tract.

### 10.2.2 Chemistry and Biochemistry

Chromium is a transition element commonly found in valence states of 0, +2, +3, and +6, with $Cr^{+3}$ being the most common form in biological systems. There are multiple radioisotopes of Cr, but most have very short half-lives and the only isotope commonly used in biology is $^{51}Cr$. This isotope decays by electron capture with a half-life of 27.7days. It emits gamma radiation and 100% percent of the emissions occur at 320.1 KEV.

Limited absorption of $Cr^{+3}$ (<5.0%) occurs in the intestine and various dietary components including starch, ascorbic acid, and zinc, apparently affect absorption. Chromium may be transported in the blood bound to transferrin, and chromium accumulates in the liver, spleen, soft tissue, and bone.[46] Human studies have found three physiological pools of Cr. A rapid turnover pool with a half-life of 0.5 days probably represents the plasma; a medium-term turnover pool has a half-life of around 13 days and a long-term pool has a biological half-life of around 190 days.[47] Urine is the primary route of Cr excretion. Chromium is considered to be an essential nutrient because of its role in potentiating insulin action.[48]

### 10.2.3 Radioactive Chromium in Human Studies

Radioactive $^{51}Cr$ is primarily used in human studies as a biological marker. The method of reinfusing erythrocytes labelled with $^{51}Cr$ is considered to be the "Gold Standard" for determining human blood, plasma, and erythrocyte volumes.[49,50] Reinfusion of $^{51}Cr$ labelled erythrocytes also has been used to study functional impairment of the reticulo-endothelial system in alcoholics, and lymph-node barrier function in normal humans.[51,52] Additionally, $^{51}Cr$ has been used as a marker of alveolar fluid volume, glomerular filtration rate, changes in intestinal permeability and fecal blood loss.[53–60] Most studies use $^{51}CrCl_3$ or $^{51}Cr$ complexed to various ligands such as ethylene diamine tetra-acetic acid (EDTA).[61]

#### *10.2.3.1 Nutritional Studies with $^{51}Cr$*

Numerous nutritional studies have used $^{51}Cr$ as a non-absorbable fecal marker[59,60,62] to determine the transit time of undigested food residue in the gastrointestinal tract. A method has been developed that estimates nutrient absorption from a single fecal sample by comparing the ratio of radioactivity of $^{51}Cr$ (given as a fecal marker) and a radioactive tracer of the nutrient of interest (administered simultaneously with the $^{51}Cr$).[35]

A limited number of human studies has used $^{51}Cr$ to study the nutritional essentiality of Cr. Sargent et al. used $^{51}Cr$ to determine differences in whole-body retention of $^{51}Cr$ in normal patients and patients with hemachromatosis.[63,64] Lim et al. studied the distribution, retention, and excretion of $^{51}Cr$ in humans by using whole-body scintillation counting, whole-body counting, and plasma counting; $^{51}Cr$ was visualized primarily in the liver, spleen, soft tissue, and bone.[65] Plasma kinetics of $^{51}Cr$ were modelled by fast ($t\frac{1}{2}$ = 0.5 to 12 hours), medium (1 to 14 days), and slow (3 to 12 months) turnover compartments.

#### *10.2.3.2 Stable Isotopes of Chromium in Human Studies*

Stable isotopes of Cr have been used for many of the same purposes as $^{51}Cr$. $^{53}Cr$ (natural abundance of 9.5%) has been used as a marker to determine red cell and total blood volume.[66,67] Similarly, $^{50}Cr$ (natural abundance of 4.35%) has been used as a marker to determine blood volume, erythrocyte volume, and erythrocyte survival.[68–70]

Nutritional functions of Cr have been studied with $^{53}Cr$. Mohamedshah et al. fed lactating women 400 μg Cr/d for three days and measured isotope abundance in urine and serum.[71] Rubin et al. fed 300 μg of $^{53}CrCl_3$ to older men in a strength-training regimen and measured urinary excretion of the isotope.[72]

## 10.3 Molybdenum

### 10.3.1 Chemistry and Biochemistry

Molybdenum (Mo) is a transition element with common valence states of +3, +4, +5, and +6. Molybdenum functions in enzymes where it is found in valence states of +4 to +6. There are eight radioisotopes and six stable isotopes of Mo. The primary radioactive isotope is the beta emitter $^{99}Mo$ with a half-life of 2.75 days. The stable isotopes of Mo and their abundances are $^{94}Mo$ (9.25%), $^{95}Mo$ (15.92), $^{96}Mo$ (16.68%), $^{97}Mo$ (9.55%), $^{98}Mo$ (24.3%), and $^{100}Mo$ (9.63%).

The majority of Mo in food is absorbed (as much as 90%); absorption occurs by an unknown mechanism. Molybdenum circulates in the blood bound to protein, accumulates in liver, kidneys, bone and skin, and is excreted through the urine. Molybdenum is considered essential because it is a co-factor in xanthine oxidase, aldehyde oxidase, and sulfite oxidase. Molybdenum is also used to activate adenylate cyclase in several organs and tissues.[73]

### 10.3.2 Radioactive Isotopes of Molybdenum in Human Studies

Radioactive $^{99}Mo$ is a contamination product of $^{99}Tc$, which is used as a biomarker for a number of medical procedures. Pharmaceuticals are often labelled with $^{99}Tc_m$ that is usually produced on-site with a commercially available

$^{99}$Mo/$^{99}$Tc generator.[74] Human study guidelines require that a dose of $^{99}$Tc contains less than 1% of contaminating $^{99}$Mo.[75] Shearer et al. measured whole-body retention of $^{99}$Mo in 14 patients administered $^{99}$Tc that had greater than 1% contaminating $^{99}$Mo and calculated the biological half-life to be 11.2 to 19.3 days.[75] The effective dose of radiation from $^{99}$Mo to the liver may be 3 to 4 times that from $^{99}$Tc.[76] A kinetic model to describe Mo retention in the human body has been developed.[77]

### 10.3.3 Stable Isotopes of Molybdenum in Human Studies

Turnlund et al. labelled kale, soybeans, and wheat with $^{100}$Mo and then fed young women a sufficient amount of each food to provide 100 µg of the isotope.[78] Isotopes in urine and feces were detected by thermal ionization mass spectrometry. Mo absorption was between 60 and 90% for the three foods. Turnlund et al. used a combination of orally consumed $^{100}$Mo and intravenously infused $^{97}$Mo to estimate the minimum daily intake of Mo needed to maintain Mo balance.[79] A similar combination of infused and fed isotopes of Mo was used to develop a compartmental model of Mo metabolism.[80]

Cantone et al. determined human fractional absorption of Mo by feeding $^{96}$Mo in an aqueous solution, feeding $^{96}$Mo in infant formula, or injecting $^{95}$Mo intravenously.[81] Isotopes were analyzed by proton nuclear activation and measurement of characteristic gamma radiation.[82] Fractional absorption was calculated to be between 0.84 and 0.98 for Mo in aqueous solution and 0.51 for Mo in infant formula.

## 10.4 Summary

Many human studies have utilized isotopes of Mn, Cr, and Mo as tracers. Manganese is studied because it is an essential nutrient and because it is potentially toxic. There is only one stable isotope of Mn; consequently, human studies that use a tracer must use radioactive Mn. A facility with a human whole-body counter also is necessary to measure the small amounts of radioactive Mn that are retained in the body. Tracers (radioactive and stable isotopes) of Cr have been used in many biological studies as markers to determine physiological pool volumes and flow rates. A few studies have used isotopes of Cr to study Cr nutrition. Radioactive isotopes of Mo are primarily encountered in medical/pharmaceutical studies where they are a contaminant of $^{99}$Tc solution. Current guidelines suggest that less than 1% of a dose of $^{99}$Tc should be $^{99}$Mo. Molybdenum is also an essential nutrient; its nutritional functions have been investigated by using a stable isotope of Mo.

## References


1. National Research Council, *Recommended Dietary Allowances*, National Academy Press, Washington, 1989.
2. Cook, D., Fahn, S., and Brait, K., Chronic manganese intoxication, *Arch. Neurol.*, 30, 59, 1974.
3. Hauser, R. et al., Manganese intoxication and chronic liver failure, *Ann. Neurol.*, 36, 871, 1994.
4. Gibson, R., Content and bioavailability of trace elements in vegetarian diets, *Am. J. Clin. Nutr.*, 59, 1223S, 1994.
5. Finley, J. and Monroe, P., Mn Absorption: the use of CACO-2 cells as a model of the intestinal epithelium, *Nutr. Biochem.*, 8, 92, 1997.
6. Finley, J.W., Manganese absorption and retention by young women is associated with serum ferritin concentration, *Am. J. Clin. Nutr.*, 70, 37, 1999.
7. Davis, C., Zech, L., and Greger, J., Manganese metabolism in rats: an improved methodology for assessing gut endogenous losses, *Proc. Soc. Exp. Biol. Med.*, 202, 103, 1993.
8. Davidsson, L. et al., Identification of transferrin as the major plasma carrier protein for manganese introduced orally or intravenously or after *in vitro* addition in the rat, *J. Nutr.*, 119, 1461, 1989.
9. Brandt, M. and Schramm, V., Mammalian manganese metabolism and manganese uptake and distribution in rat hepatocytes, in *Manganese in Metabolism and Enzyme Function*, Schramm, V. and Wedler, F., Eds., Academic Press, Orlando, 1986, 3–16.
10. Classen, C., Biliary excretion of manganese in rats, rabbits and dogs, *Toxicol. Appl. Pharma.*, 29, 458, 1974.
11. Wedler, F., Biochemical and nutritional role of manganese: an overview, in *Manganese in Health and Disease*, Klimis-Tavantzis, D., Ed., CRC Press, Boca Raton, 1994, 1–38.
12. Dastur, D., Manghani, D., and Raghavendran, K., Distribution and fat of $^{54}$Mn in the monkey: studies of different parts of the central nervous system and other organs, *J. Clin. Invest.*, 50, 9, 1971.
13. King, B. et al., Effect of lactose, copper and iron on manganese retention and tissue distribution in rats fed dextrose-casein diets, *J. Anim. Sci.*, 50, 452, 1980.
14. Schramm, V. and Brandt, M., The manganese (II) economy of rat hepatocytes, *Fed. Proc.*, 45, 2817, 1986.
15. Finley, J., Manganese uptake and release by cultured human hepato-carcinoma (Hep-G2) cells, *Biol. Trace Elem. Res.*, 64, 101, 1998.
16. Rabin, O. et al., Rapid brain uptake of manganese II across the blood-brain barrier, *J. Neurochem.*, 61, 509, 1993.
17. Nishida, M. et al., A binding profile of manganese to the nucleus of rat liver cells, and manganese induced aberrations in thyroid hormone content and RNA synthesis in the nucleus, *Biochem, Inter.*, 27, 209, 1992.
18. Keefer, R., Barak, A., and Boyett, J., Binding of manganese and transferrin in rat serum, *Biochim. Biophys. Acta*, 221, 390, 1970.
19. Finley, J.W. and Monroe, P., The use of CACO-2 cells as a model of the intestinal epithelia, *J. Nutr. Biochem.*, 8, 92, 1997.

20. Strause, L. et al., The oral assimilation of radiomanganese by the mouse, *Biol. Trace Elem. Res.*, 7, 75, 1985.
21. Weigand, E., Helbig, U., and Kirchgessner, M., Radioisotope-dilution technique for determining endogenous manganese in feces of the growing rat, *Biol. Trace Elem. Res.*, 10, 281, 1986.
22. Lee, D. and Johnson, P., Factors affecting absorption and excretion of $^{54}$Mn in rats, *J. Nutr.*, 118, 1509, 1988.
23. Lee, D. and Johnson, P., $^{54}$Mn absorption and excretion in rats fed soy protein and casein diets (42852), *Proc. Soc. Exp. Biol. Med.*, 190, 211, 1989.
24. Lee, D., Korynta, E., and Johnson, P., Effects of sex and age on manganese metabolism in rats, *Nutr. Rev.*, 10, 1005, 1990.
25. Davis, C.D. et al., Effect of biliary ligation on manganese accumulation in rat brain, *Biol. Trace Elem. Res.*, 64, 61, 1998.
26. Roughead, Z.K., Hunt, J., and Finley, J., Hepatic iron is elevated, but mucosal uptake and whole body retention of dietary iron and manganese are not altered in young $\beta_2$-microglobulin knockout mice, *Biol. Trace Elem. Res.*, in press.
27. Diez-Ewald, M., Weintraub, L., and Crosby, W., Interrelationship of iron and manganese metabolism, *Proc. Soc. Exp. Biol. Med.*, 129, 448, 1968.
28. Thomson, A.B.R. et al., Interrelation of intestinal transport system for manganese and iron, *J. Lab. Clin. Med.*, 78, 642, 1971.
29. Cikrt, M. and Tichy, M., Polyacrylamide gel disc electrophoresis of rat bile after intravenous administration of 52MnC12, 64CuC12, 203HgC12, and 210Pb(NO3)2, *Special.*, 15, 383, 1972.
30. Borg, D. and Cotzias, G., Manganese metabolism in man: rapid exchange of Mn56 with tissue as demonstrated by blood clearance and liver uptake, *J. Clin. Invest.*, 37, 1269, 1958.
31. Sheppard, C. et al., Studies of the distribution of intravenously administered colloidal sols of manganese dioxide and gold in human being and dogs using radioactive isotopes, *J. Lab. Clin. Med.*, 32, 274, 1947.
32. Cotzias, G. et al., Chronic manganese poisoning: clearance of tissue manganese concentrations with persistence of the neurological picture, *Neurology*, 18, 376, 1968.
33. Mena, I. et al., Chronic manganese poisoning: individual susceptibility and absorption of iron, *Neurology*, 19, 1000, 1969.
34. Mahoney, J. and Small, W., Studies on manganese III. The biological half-life of radiomanganese in man and factors which affect this half-life, *J. Clin. Invest.*, 47, 643, 1968.
35. Davidsson, L. et al., Manganese retention in man: a method for estimating manganese absorption in man, *Am. J. Clin. Nutr.*, 49, 170, 1989.
36. Davidsson, L. et al., Intrinsic and extrinsic labeling for studies of manganese absorption in humans, *J. Nutr.*, 118, 1517, 1988.
37. Johnson, P.E., Lykken, G.I., and Korynta, E.D., Absorption and biological half-life of intrinsic and extrinsic $^{54}$Mn tracers from foods of plant origin, *J. Nutr.*, 121, 711, 1991.
38. Johnson, P. and Lykken, G., Manganese and calcium absorption and balance in young women fed diets with varying amounts of manganese and calcium, *J. Trace Elem. Exp. Med.*, 4, 19, 1991.
39. Finley, J. and Johnson, P., Sex affects manganese absorption and retention by humans from a diet adequate in manganese, *Am. J. Clin. Nutr.*, 60, 949, 1994.

40. Leggett, R., A retention-excretion model for Ameridium in humans, *Health Phys.*, 62, 288, 1992.
41. Wastney, M., Zinc absorption in humans determined using *in vivo* tracer studies and kinetic analysis, in Mineral Absorption in the Monogastric GI Track, *Adv. Exp. Med. Biol.*, 249, 3, 1989.
42. Dunn, M., Green, M., and Leach, R., Kinetics of copper metabolism in rats: a compartmental model, *Am. J. Physiol.*, 261, E115, 1991.
43. Swanson, C.A. et al., Human [$^{74}$Se]selenomethionine metabolism: a kinetic model, *Am. J. Clin. Nutr.*, 54, 917, 1991.
44. Finley, J.W., Johnson, P.E., and Johnson, L.K., A compartmental model of manganese metabolism in rats, *FASEB J.*, 6, A1947, 1992.
45. Finley, J. et al., A surgical model for determination of true absorption and biliary excretion of manganese in conscious swine fed commercial diets, *J. Nutr.*, 127, 2334, 1997.
46. Stoecker, B.J., Chromium, in *Present Knowledge in Nutrition*, 7$^{th}$ ed, Ziegler, E.E. and Filer, Jr., L.J., Eds., ILSI Press, Washington, 1996, 344–352.
47. Vuori, E. and Kumpulainen, J., A new low level of chromium in human liver and spleen, *Trace Elem. Med.*, 4, 91, 1987.
48. Mertz, W., Chromium in human nutrition: a review, *J. Nutr.*, 123, 626, 1993.
49. Fairbankis, V.F. et al., Measurement of blood volume and red cell mass: re-examination of 51Cr and 125I methods [see comments], *Blood Cells Mol. Dis.*, 22, 169, 1996.
50. Bernard, P.J., Measurement of red-cell and plasma volumes, *Nouv. Rev. Fr. Hematol.*, 36, 155, 1994.
51. Cohen, A.J. and Minuk, G.Y., The effect of acute alcohol ingestion on Fc-receptor-mediated clearance of IgG-tagged erythrocytes by the reticuloendothelial system in humans, *Alcohol*, 10, 181, 1993.
52. Peters, P.E., The function of the lymph node barrier in humans. Studies with Cr51-labeled erythrocytes, *Fortschr. Geb. Rontgenstr. Nuklearmed.*, 113, 83, 1970.
53. Von Wichert, P. et al., Bronchoalveolar lavage, quantitation of intraalveolar fluid? [see comments], *Am. Rev. Respir. Dis.*, 147, 148, 1993.
54. Soroka, N. et al., A Comparison of a vegetable-based (soya) and an animal-based low-protein diet in predialysis chronic renal failure patients, *Nephrology*, 79, 173, 1998.
55. Bouhanick, B. et al., Relationship between fat intake and glomerular filtration rate in normotensive insulin-dependent diabetic patients, *Diabete Metab.*, 21, 168, 1995.
56. Sawamura, R. et al., Increased intestinal permeability to 51 Cr-EDTA among children with persistent diarrhea, *Arq. Gastroenterol.*, 34, 55, 1997.
57. Sobotka, L. et al., Inulin as the soluble fiber in liquid enteral nutrition, *Nutrition*, 13, 21, 1997.
58. Oman, H. et al., Detection of naproxen-induced intestinal permeability change may be facilitated by adding a standardized meal but not by forming marker ratios, *Scand. J. Gastroenterol.*, 31, 1182, 1996.
59. Wood, R.J. and Zheng, J.J., Milk consumption and zinc retention in postmenopausal women, *J. Nutr.*, 120, 398, 1990.
60. Bassett, M.L. and Goulston, K.J., False positive and negative hemoccult reactions on a normal diet and effect of diet restriction, *Aust. N. Z. J. Med.*, 10, 1, 1980.
61. Bjarnason, I. et al., Glucose and citrate reduce the permeability changes caused by indomethacin in humans, *Gastroenterology*, 102, 1546, 1992.

62. Daher, G.C. et al., Olestra ingestion and dietary fat absorption in humans, *J. Nutr.*, 127, 1694S, 1997.
63. Sargent, III., Lim, T.H., and Jenson, R.L., Reduced chromium retention in patients with hemochromatosis, a possible basis of hemochromatotic diabetes, *Metabolism*, 28, 70, 1979.
64. Sargent, III and Stauffer, H., Whole-body counting of retention of $^{67}$Cu, $^{32}$P and $^{51}$Cr in man, *Int. J. Nucl. Med. Biol.*, 6, 17, 1979.
65. Lim, T.H., Sargent, III, and Kusubov, N., Kinetics of trace element chromium(III) in the human body, *Am. J. Physiol.*, 244, R445, 1983.
66. Silver, H.M. et al., Red cell volume determination using a stable isotope of chromium, *J. Soc. Gynecol. Investig.*, 4, 254, 1997.
67. Veillon, C. et al., Measurement of blood volume with an enriched stable isotope of chromium ($^{53}$Cr) and isotope dilution by combined gas chromatography-mass spectrometry, *Clin. Chem.*, 40, 71, 1994.
68. Yamabayashi, H. et al., Blood volume measurement of newborn using stable isotope $^{50}$Cr, *Radioisotopes*, 34, 144, 1985.
69. Faxelius, G. et al., Red cell volume measurements and acute blood loss in high-risk newborn infants, *J. Pediatr.*, 90, 273, 1977.
70. Glomski, C.A., Pillay, K.K., and Macdougall, L.G., Erythrocyte survival in children as studied by labeling with stable $^{50}$Cr, *Am. J. Dis. Child*, 130, 1228, 1976.
71. Mohamedshah, F.Y. et al., Distribution of a stable isotope of chromium ($^{53}$Cr) in serum, urine, and breast milk in lactating women, *Am. J. Clin. Nutr.*, 67, 1250, 1998.
72. Rubin, M.A. et al., Acute and chronic resistive exercise increase urinary chromium excretion in men as measured with an enriched chromium stable isotope, *J. Nutr.*, 128, 73, 1998.
73. Nielsen, F.H., Other Trace Elements, in *Present Knowledge in Nutrition*, 7th ed., Ziegler, E.E. and Filer, Jr., L.J., Eds., ILSI Press, Washington, 1996, 353–377.
74. Hjelstuen, O.K., Technetium-99m chelators in nuclear medicine, a review, *Analyst*, 120, 863, 1995.
75. Shearer, D.R. et al., Radiation dose from radiopharmaceuticals contaminated with molybdenum-99, *J. Nucl. Med.*, 29, 695, 1988.
76. Nagaratnam, A., Kaul, A., and Roedler, H.D., Radiation dose to various age groups from radionuclide impurities in 99mTc-pertechnetate (fission product $^{99}$Mo generator) radiopharmaceutical preparations, *Eur. J. Nucl. Med.*, 14, 331, 1988.
77. Giussani, A. et al., Internal dose for ingestion of molybdenum radionuclides based on a revised biokinetic model, *Health Phys.*, 78, 46, 2000.
78. Turnlund, J.R. et al., Molybdenum absorption and utilization in humans from soy and kale intrinsically labeled with stable isotopes of molybdenum, *Am. J. Clin. Nutr.*, 69, 1217, 1999.
79. Turnlund, J. R., Keyes, W.R., and Peiffer, G.L., Molybdenum absorption, excretion, and retention studied with stable isotopes in young men at five intakes of dietary molybdenum, *Am. J. Clin. Nutr.*, 62, 790, 1995.
80. Thompson, K.H. and Turnlund, J.R., Kinetic model of molybdenum metabolism developed from dual stable isotope excretion in men consuming a low molybdenum diet, *J. Nutr.*, 126, 963, 1996.
81. Cantone, M.C. et al., A methodology for biokinetic studies using stable isotopes: results of repeated molybdenum investigations on a healthy volunteer, *Appl. Radiat. Isot.*, 48, 333, 1997.
82. Cantone, M.C. et al., Proton activation analysis of stable isotopes for a molybdenum biokinetics study in humans, *Med. Phys.*, 22, 1293, 1995.

# 11

# *Trace-element Studies in Infants and Pregnant or Lactating Women*

**Lena Davidsson**

**CONTENTS**

11.1 Introduction .......... 167
11.2 Iron .......... 170
 11.2.1 Methodology .......... 170
 11.2.2 Erythrocyte Incorporation and Iron Absorption .......... 173
  11.2.2.1 Studies in Premature Infants .......... 173
  11.2.2.2 Studies in Term Infants .......... 174
   11.2.2.2.1 Human Milk and Infant Formula .......... 174
   11.2.2.2.2 Complementary Foods .......... 176
   11.2.2.2.3 Iron Supplements .......... 177
  11.2.2.3 Studies in Pregnant Women .......... 177
11.3 Zinc .......... 178
11.4 Zinc and Copper .......... 180
11.5 Selenium .......... 181
11.6 Chromium .......... 182
11.7 Conclusion .......... 183
References .......... 183

## 11.1 Introduction

The development of stable-isotope techniques to study metabolism of trace elements has provided much-needed tools to implement studies in vulnerable segments of the population such as infants and pregnant or lactating women. The major advantage of stable-isotope techniques is that they can be used without introducing any risk to the study subjects. Longitudinal studies to investigate changes in trace element metabolism, for example, changes in

0-8493-8730-2/00/$0.00+$.50

iron metabolism after treatment with erythropoietin in preterm infants or trace element metabolism during pregnancy and lactation, have been made with repeated administrations of stable-isotope labels.[1,2] Furthermore, several stable-isotope labels can be administered and monitored simultaneously, to study different trace elements, different compounds, or different routes of administration.[3–12] However, so far only a limited number of studies have been reported on the application of these techniques in infants and pregnant or lactating women.

The high cost of stable-isotope labels, the cost of analyses, and access to analytical equipment, are major drawbacks of these techniques. Doses of stable-isotope labels to be used in human studies of trace-element metabolism depend on a number of factors, including analytical precision and expected enrichment in biological samples. In addition, body size and blood volume are important for dose calculations in some study protocols, such as for studies of erythrocyte incorporation of iron stable isotopes.[13] Consequently, smaller doses of administered stable-isotope labels can be used in infants as compared to older children or adults. Due to the high cost of studies based on stable-isotope techniques, the development of and adherence to strictly standardized study protocols are essential. Nutritional studies in vulnerable groups, in particular in infants, are complicated since compliance with study protocols can be difficult to achieve. Strictly standardized study protocols including fasting, intake of labelled test meals at fixed time points, no food or fluid intake between test meals, etc., feasible in studies in healthy adults, are sometimes difficult to follow in infant studies. However, although more difficult to implement, the use of strictly standardized study protocols facilitates the interpretation of results. Due to large interindividual variation in trace element metabolism, crossover study designs with paired comparisons should be used whenever possible, for example, to evaluate the influence of dietary factors on trace-element bioavailability. The development of study protocols should be based on careful considerations, including the definition of the study population (age, previous diet, etc.) and the size and composition of labelled test meals, as well as the dose of stable-isotope labels.

The size of the labelled test meals has often been chosen to be small, or very small, in infant studies to assure complete intake of the serving and the entire dose of stable-isotope label(s). The relevance of the results in relation to normal feeding practices is therefore not always obvious. To some extent, these problems can be overcome by using study protocols more closely resembling normal feeding patterns by feeding labelled test meals at each feeding over approximately 24 hours, or by administration of labelled foods over several days under standardized conditions.[8,10,14–18] Obviously, such study protocols are more complicated to follow and more time-consuming for the study subjects as well as for the investigators. For obvious ethical and practical reasons, infants cannot be kept without food for more than a few hours. However, in order to standardize the intake of stable-isotope labels, administrations should preferably be done 2 to 3 hours after the last feeding to minimize interference with residual food components in the stomach, and no additional feedings should be given during the following 2 to 3 hours.[19,20] The influence of unrestricted intake of food or fluid immediately after intake of stable-isotope labels is very difficult to evaluate due to differences in volume and total trace element content in the mixture of labelled and unlabelled food(s) and should be avoided. In addition, practical problems such as regurgitation of ingested food, which is not uncommon in infants, is a major concern when feeding babies labelled test meals. Small volumes of iron solutions, labelled with iron stable isotopes, have been delivered directly into the back of the oral cavity by syringe to facilitate administration in some studies.[14,19,20] Furthermore, frequent episodes of infections in otherwise healthy infants can easily modify appetite and willingness to comply with study protocols. On the other hand, the monotony of standardized diets during metabolic studies, which can be a problem when study subjects are used to a mixed, self-selected diet, is less a concern in infants, in particular during the first 4 to 6 months of life when human milk or infant formula is the only food consumed.

The total content of trace element(s) and ratios to other nutrients in labelled test meals can be substantially modified after addition of stable-isotope labels, in particular in foods with low trace-mineral content. For example, the concentrations of trace elements have been increased significantly in human milk after labelling with stable isotopes.[5,14,21] The extent to which the increased trace-element content will influence the results is difficult to evaluate. Administered stable-isotope doses were relatively high, and test-meal sizes small, in some of the studies referred to in this review; therefore, the results are difficult to interpret in relation to practical infant nutrition. A close collaboration with analytical chemists responsible for dose calculations and sample analysis is essential to explore possibilities to decrease doses of stable-isotope labels without jeopardizing the enrichment of biological samples. Doses of stable-isotope labels have varied considerably between studies, depending on study design and precision of the analytical techniques used in different studies and, therefore, have not been commented on in detail in this review. Stable-isotope labels, usually as water-soluble compounds such as sulfate, chloride or citrate, have been added by extrinsic labelling technique in most studies in infants and pregnant or lactating women, but exceptions are indicated in the text. The labelled compounds are specified when of special interest, for example, when different compounds of iron and selenium have been compared.

The labor-intense study protocols required for complete collections of excreta and the reluctance to use intravenous injections of stable-isotope labels in vulnerable population groups are major limitations to the wider application of stable-isotope techniques in trace-element studies. The extended periods of complete fecal collections needed to recuperate all non-absorbed stable-isotope labels are major drawbacks in studies based on fecal monitoring technique. For example, in the studies reported by Swanson et al. and Turnlund et al., the appearance of $^{70}Zn$ and $^{65}Cu$ was monitored for 12 days in pregnant women.[3,4] Infants have shorter gastrointestinal passage time as compared with adults and, therefore, a shorter fecal collection period can be applied. In most infant studies, complete fecal collections over 72 to 96 hours have been made using fecal markers such as carmine red or brilliant blue to determine the start and endpoint of the collection. Fecal material has been collected for longer periods in studies where estimates of re-excretion of absorbed trace elements have been included, e.g., during 21 days in a study of re-excretion of absorbed $^{70}Zn$.[22] Collections of fecal material during infancy are, to some extent, simplified since babies in industrialized countries wear diapers. However, for complete collections of feces, in particular when complete collections of urine are also required, the use of metabolic beds is the best option.[15,23,24] Metabolic beds are only available in a few centers, however, and alternative methodologies for the collection of feces have been developed, especially for studies in healthy, non-hospitalized infants. For example, the use of trace-element-free diapers with diaper liners, portable collection seats with a removable plastic liner bag, or collection of feces on ashless filter paper placed in diapers has been described.[22,25,26] Collections of urine are complicated in non-hospitalized infants but can be made with the use of adhesive urine collection bags.[26]

Methodology based on tissue incorporation, for example, erythrocyte incorporation of iron stable-isotope labels, has obvious advantages since no collections of feces or urine are needed. The only sampling required is a small volume of whole blood, venous or capillary, 14 days after intake of stable-isotope labels. The recent development of this technique is a major focus in the area of trace element studies in infants and has provided new, important information about iron bioavailability from infant diets and iron metabolism during early life.

## 11.2 Iron

### 11.2.1 Methodology

The first reports on erythrocyte incorporation of $^{58}Fe$ as a new, promising technique to study iron availability in infants were published in 1986 and 1988.[27,28] In this first infant study, 1.44 mg $^{58}Fe$ (as ferrous sulfate together with 84 mg sodium ascorbate) was given between formula feedings to nine 126-day-old infants. The percentage of $^{58}Fe$ entering the circulation was based on mean $^{58}Fe$ in blood samples drawn at 140-, 168-, and 196-days-of-age. The large individual variation in $^{58}Fe$ incorporation found in this study (range 3.2 to 16.0%; geometric mean 7.9%) was inversely correlated with serum ferritin values and indicated that this technique is suitable for within-subject comparisons of iron bioavailability from foods, but less useful for group comparisons. The later development of a double-isotope technique, using administrations of $^{57}Fe$ and $^{58}Fe$, provided the possibility to study iron bioavailability from two test meals administered on consecutive days.[13] In the study by Kastenmayer et al., extrinsically labelled infant formula was administered on 4 consecutive days (210 ml formula with 2.5 mg total iron/test meal) to healthy infants 13 to 25 weeks old.[13] Erythrocyte incorporation was measured by thermal ionization mass spectrometry 14 days after intake, resulting in geometric mean iron absorption of 6.7 and 6.6% for $^{57}Fe$ and $^{58}Fe$, respectively. Total doses of stable-isotope labels were about 5 mg $^{57}Fe$ and 1.2 mg $^{58}Fe$ per infant.

The authors give detailed information about calculations and data treatment.[13,27,28] Calculations of circulating total iron are based on information about blood volume, hemoglobin concentration, and the iron content of hemoglobin (3.47 mg/g). Direct measurements of blood volume are not available in any of the infant studies published to date and the estimate of blood volume (65 ml blood/kg), as used by Kastenmayer et al., Janghorbani et al., and Fomon et al., has been used in most studies of healthy infants.[13,27,28] For studies in preterm infants, 80 to 85 ml blood/kg body weight is usually assumed.[1,11,12,29]

When results based on erythrocyte incorporation are presented as absorption data in infants, as in the study by Kastenmayer et al., 90% of absorbed iron has often been assumed to be incorporated into erythrocytes.[13] In adults, 80 to 100% of retained iron is incorporated into red blood cells 14 days after intake of radioisotopes of iron (reviewed by Fomon et al.).[30] However, only very limited information on the rate of erythrocyte incorporation of newly absorbed iron in infants is available. Studies in preterm infants, and recently also in term infants, have reported that erythrocyte incorporation is lower than in adults and thus does not support the assumption that red blood cell incorporation is a valid surrogate for iron absorption early in life. Iron absorption based on fecal monitoring has been compared with erythrocyte incorporation of iron stable-isotope labels in some of these studies. A study in 11 preterm infants demonstrated considerable differences between iron absorption from $^{58}Fe$ (228 μg/kg fed by nasogastric tube with 10 mg ascorbic acid/kg), based on 7-day fecal collection (41.6±17.6%) and erythrocyte incorporation on day 15 (12.0±9.6%).[29] In a recent study by Widness et al., iron absorption based on fecal excretion of $^{58}Fe$ collected for 10 days after intake of $^{58}Fe$ (0.45 mg $^{58}Fe$/kg given with 5.55 mg iron with normal isotopic composition and 30 mg ascorbic acid/kg) was compared with erythrocyte incorporation after 2 weeks in preterm infants treated with erythropoietin or in untreated infants.[1] Fractional mean absorption was similar (30 to 34%) in both groups and at the two time points studied (1 and 4 weeks after treatment). However, erythrocyte incorporation after 2 weeks was much lower than iron absorption, and significantly higher in the treated group compared with placebo (4.4±1.6% vs 2.0±1.4%) one week after treatment. No difference was observed, however, between the two groups at 4 weeks (3.8±1.6% vs 5.5±2.7%).

Erythrocyte incorporation can be determined after administration of an iron stable-isotope label intravenously, assuming that the availability of iron for erythropoiesis is the same after intravenous administration as from iron absorbed from an oral dose. Two studies in premature infants report erythrocyte incorporation after intravenous administration of iron stable-isotope labels. In the study by Zlotkin et al., a substantial amount of $^{57}Fe$ (0.09 to 0.13 mg $^{57}Fe$ as ferrous sulfate; target 0.15 mg $^{57}Fe$/kg) was given intravenously to six very-low-birth-weight infants (720 to 1320 g at baseline).[11] Erythrocyte incorporation after 14 days was in the range 4.5 to 34.3% of the dose. However, interpretation of the data is complicated by the relatively high dose of administered iron and by the numerous blood transfusions received by the study infants. In the study by McDonald et al., 13 premature infants (mean weight 1599 g at study entry) were given 0.2 mg $^{58}Fe$ intravenously (as $^{58}Fe$-citrate) and erythrocyte incorporation was determined 14 days later.[12] Erythrocyte $^{58}Fe$ incorporation was much higher than in the study by Zlotkin et al.; geometric mean was 67.7% (range 51.8 to 84.8%). Thus, the results based on erythrocyte incorporation after intravenous administration of iron-stable isotopes to premature infants are not conclusive and difficult to interpret. No comparable information is available for term infants.

Recently, results from a study in term infants reported significant differences in iron retention and erythrocyte incorporation after intake of 0.8 mg $^{58}Fe$ between feedings.[30] Geometric mean $^{58}Fe$ retention, based on 11-day fecal collections, and erythrocyte incorporation 14 days after intake were 26.9% and 5.2%, in nine young infants (20 to 69 days) and 32.5% and 12.5%, respectively, in nine older infants (165 to 215 days).[30] The authors conclude that infants incorporate far less than 80% of retained isotope into erythrocytes. Thus, estimating iron retention based on the assumption that 80 to 100% of absorbed iron is promptly incorporated into erythrocytes will result in a several-fold underestimate of retention.

In addition, a recent study by Fomon et al. reports new data on the influence of age and dietary iron intake on erythrocyte incorporation of iron.[31] In this study, 0.45 mg iron (0.4 mg $^{58}Fe$ plus 10 mg ascorbic acid) was given twice on the same day to twenty-one 56-day-old and twenty 168-day-old infants. All infants were fed a low-iron formula (0.3 mg/100 kcal) before, and until 5 hours after, completed administration of $^{58}Fe$. Half of the infants per age group were fed a formula high in iron (1.8 mg/100 kcal) and the other infants continued consuming the low-iron formula during the rest of the study. Erythrocyte incorporation of $^{58}Fe$ increased from 14 to 28 days postdosing, remained relatively constant until day 84, and then declined between days 84 to 112. Red blood cell incorporation in infants consuming high-iron formula was significantly higher in the older infants (geometric mean 15.5%) compared with the 56-day-old infants (geometric mean 9.2%). In the 56-day-old infants, red blood cell incorporation of $^{58}Fe$ was significantly higher in the infants consuming low-iron formula (geometric mean 15.4%) than in infants fed high-iron formula (geometric mean 9.2%), but not in 168-day-old infants (geometric mean 19.8% and 15.5%). Thus, age had a significant effect on erythrocyte incorporation of $^{58}Fe$ and the level of iron intake after intake of the iron stable-isotope label influenced erythrocyte incorporation in younger infants. Furthermore, data from this study indicate that the life span of erythrocytes during infancy is less than 112 days.[31]

The incorporation rate of newly absorbed iron into red blood cells is an important methodological issue and is of concern when evaluating results from stable-isotope studies as quantities of absorbed iron. However, in studies in which relative bioavailability is evaluated, e.g., the effect of dietary enhancers and inhibitors or the difference in physiological state on iron absorption, the incorporation rate does not influence the data when a constant factor is used for re-calculations of erythrocyte incorporation to fractional absorption.

In the following review, iron absorption was based on erythrocyte incorporation day 14, re-calculated to absorption (sometimes bioavailability) by assuming 90% erythrocyte incorporation of newly absorbed iron. Administered doses of stable-isotope labels in different studies are indicated in the methodological section as well as those of special interest, for example, when administered to premature infants or when added to infant foods with very low-iron content such as human milk. Results are given as geometric mean and range or as arithmetic mean ±SD. Iron isotope labels have usually been administered as soluble iron compounds, mostly as ferrous sulfate. The administration of other labelled iron compounds is indicated in the text. "Iron" in this review refers to non-heme iron since bioavailability of heme iron has not been studied by stable isotope technique in infants or pregnant or lactating women.

## 11.2.2 Erythrocyte Incorporation and Iron Absorption

### *11.2.2.1 Studies in Premature Infants*

In addition to the information on erythrocyte incorporation rate after intravenous injection of iron stable-isotope labels, papers by Zlotkin et al. and McDonald et al. also report on erythrocyte incorporation and iron absorption from iron stable-isotope labels administered orally.[11,12] Zlotkin et al. gave 0.48 to 1.5 mg $^{58}$Fe/kg together with 42 mg ascorbic acid by nasogastric feeding tube, separate from feedings.[11] Red blood cell incorporation was in the range 1.7 to 8.0%, resulting in 26.3±13.0% absorption after correcting the values for $^{57}$Fe incorporation. McDonald and colleagues fed 0.7 mg $^{57}$Fe mixed in two feedings of low-iron-containing premature infant formula and, on the next day, 2.0 mg $^{54}$Fe were given with a multivitamin supplement containing 35 mg ascorbic acid.[12] The supplement was given when less than 20% of a feeding remained in the stomach. All doses were given by orogastric feeding tube. Geometric mean iron erythrocyte incorporation was significantly higher from the labelled vitamin supplement (9.0%, range 4.1 to 16.3%) compared with infant formula (7.0%, range 1.8 to 13.7%). Corrected values resulted in iron absorption of 14.6±6.5% and 11.4±4.4% from the supplement and formula, respectively.

Other studies in premature infants have evaluated utilization of supplemental iron in premature infants fed human milk and the influence of zinc on erythrocyte incorporation of iron. Moody et al. gave 13 infants (1571±426 g at the start of the study) $^{57}$Fe (2 mg/kg) mixed with fortified human milk and fed in eight aliquots by intermittent bolus over 24 hours.[16] On the next day, $^{54}Fe$ (2 mg/kg) was given by orogastric gavage 1.5 hours after a feeding. Erythrocyte incorporation of iron stable-isotope labels was similar from the two methods of administration: 4.7±2.5% and 4.6±1.5%, respectively. A significant correlation between red blood cell incorporation of $^{57}Fe$ and the reticulocyte count was observed in this study as well as by McDonald et al.[12] In the study by Friel et al., infants (clinically stable and fed orally) received high (1200 μg/kg) or low (300 μg/kg) doses of zinc (zinc sulfate) together with 300 μg/kg $^{58}Fe$ as ferric chloride and 10 mg ascorbic acid/kg.[32] The doses were given between feeds to five infants, using a crossover study design with 2 weeks between administrations. The higher dose of zinc decreased red blood cell incorporation of $^{58}Fe$ significantly (geometric mean 3.6%) compared to the lower zinc dose (geometric mean 7.5%). However, when the doses of zinc and iron were administered with usual feeds (human milk or formula) to nine infants, $^{58}Fe$ incorporation was not influenced by the zinc content. Geometric mean erythrocyte incorporation was 6.7% and 7.0% from feedings containing high and low zinc contents, respectively.

#### ***11.2.2.2 Studies in Term Infants***

##### *11.2.2.2.1 Human Milk and Infant Formula*

The influence of lactoferrin on iron absorption in breast-fed infants was investigated by feeding human milk extrinsically labelled with $^{58}Fe$.[14] Eight infants (2- to 10-months-old) were fed similar quantities of milk expressed by their mothers (700 to 1000 g/batch labelled with 0.5 mg $^{58}Fe$ [0.54 mg total iron]) in a crossover study design. Human milk used in one part of the study contained the native lactoferrin, while milk used in the other part of the study had been treated with heparin Sepharose to remove more than 97% lactoferrin. No other foods or fluids were fed until the total quantity of labelled human milk had been consumed. A water solution (reference dose) with 3.0 mg iron (2.86 mg $^{57}Fe$ and 123 mg sodium ascorbate) was administered by syringe directly into the infant's mouth (at least 2 hours after the last feeding) on the day after completed intake of labelled human milk. Fractional iron absorption was significantly lower from untreated milk (geometric mean 11.8%, range 3.4 to 37.4%) as compared to lactoferrin-free human milk (geometric mean 19.8%, range 8.4 to 72.8%). Geometric mean absorption from the reference dose was 24.3% and 22.4% when given after intake of labelled human milk and after completed intake of lactoferrin-free human milk, respectively. Thus, these results do not support a role for lactoferrin in increasing iron absorption from human milk. Furthermore, iron availability from extrinsically labelled human milk was found to be relatively low in this study. As already discussed in the methodological section, the increase in total iron content (4- to 13-fold) is not negligible and could have negatively influenced fractional erythrocyte incorporation of $^{58}Fe$. However, the major aim of this study was to evaluate the influence of lactoferrin on iron absorption and for this purpose the dose of iron stable isotope was not a major concern.

An earlier study reported no difference in iron absorption in newborn infants after intake of infant formula with added bovine lactoferrin (44.4±25.8%, n = 13) or ferric chloride (46.2±23.9%, n = 16).[33] Iron absorption was based on fecal excretion of $^{58}Fe$ over 3 days after intake of the labelled test meals. Methodological problems related to incomplete fecal collections, partly due to constipation in some of the infants, probably influenced the data and contributed to the very wide variation in iron absorption.

Recently, Abrams et al. reported geometric mean iron absorption of 14.8% (range 1.5 to 57.2%) from extrinsically labelled human milk in 5- to 7-month-old infants.[5] Human milk (250 ml) was labelled with 180 μg $^{58}Fe$ and fed over three meals. In addition, 1.0 mg $^{57}Fe$ was given via syringe into the infant's mouth, at least 2 hours after the last meal. Geometric mean fractional absorption of $^{57}Fe$ was 11.0% (range 1.5 to 45.7%). For unknown reasons, blood volume was estimated by using 80 ml/kg in this study. When absorption data were re-calculated based on an estimated blood volume of 65 ml per kilogram body weight, geometric mean iron absorption was 12.8% from labelled human milk and 9.5% from $^{57}Fe$ given without food. The results from this study are difficult to evaluate since infants were fed unrestricted quantities of unlabelled human milk and beikost immediately after intake of labelled human milk.

Furthermore, erythrocyte incorporation of iron-stable isotopes has been used to evaluate differences between breast-fed and formula-fed infants as well as the influence of dietary inhibitors and enhancers in soy formula and the effect of iron concentration in milk-based infant formula on iron bioavailability in healthy term infants. Significant differences in erythrocyte incorporation in 14 breast-fed and 20 formula-fed 56-day-old infants were reported by Fomon et al.[19] Infants were given daily doses of 0.6 to 1.0 mg iron (0.5 to 0.6 mg $^{58}Fe$, fed with 10 mg ascorbic acid) on three consecutive days between feedings. When measured at 70-days of age, geometric mean erythrocyte incorporation was 16.0% (range 4.5 to 45.8%) in breast-fed infants and 6.0% (range 2.4 to 18.7%) in formula-fed infants. The authors suggested that, even between feedings, components from infant formula present in the stomach inhibited iron absorption in this study. Repeated analyses of blood drawn at 112-days of age demonstrated that the geometric mean erythrocyte incorporation of $^{58}Fe$ had increased significantly to 19.1% in breast-fed infants (n = 12) but not in formula-fed babies. Thus, some $^{58}Fe$ was apparently absorbed and stored for later incorporation into newly formed erythrocytes.

Erythrocyte incorporation of iron from partially (83%) dephytinized soy formula was demonstrated to be significantly increased as compared to soy formula with the native content of phytic acid (geometric mean 6.8% vs 5.5%, n = 10).[18] A more pronounced effect was found after complete degradation of phytic acid in soy formula; geometric mean erythrocyte incorporation in ten infants increased from 3.9% (native phytic acid) to 8.7% (100% dephytinized). A statistically significant enhancing effect was also demonstrated after doubling the ascorbic acid molar ratio relative to iron in soy formula from 2.1 to 4.2; geometric mean iron erythrocyte incorporation increased from 5.9

to 9.6% (n = 10).[18] Two different concentrations of fortification iron were evaluated in infants fed a milk-based infant formula with 8 mg/l or 12 mg/l of iron from 112 days of age until 196 days of age.[34] $^{58}$Fe-labelled formula (0.24 l) was fed on three consecutive days to 154-day-old infants (26 and 19 infants in the two groups; 8 mg/l or 12 mg/l). Geometric mean erythrocyte incorporation after 14 days was 3.75% and 2.29% from the formulas, with 8 or 12 mg iron/l, respectively. In this study, the quantity of iron incorporated into erythrocytes was calculated by multiplying the fractional erythrocyte incorporation of $^{58}$Fe by the quantity of total iron consumed for each infant (average for the 3 days on which the labelled test meals were fed). No statistically significant difference was found between the two groups; geometric mean incorporation of iron was 0.285 mg and 0.268 mg (measured at 168 days of age) for infants consuming formula with 8 or 12 mg iron per liter respectively. The influence of adding rice cereal to infant formula on iron absorption was evaluated in nine infants (mean age 4.1 months).[35] A small volume of infant formula (30 ml) was fed with or without small amounts of added rice cereal (6.5 g/dl), followed by intake of $^{57}$Fe or $^{58}$Fe and *ad libitum* intake of unlabelled formula. A double stable-isotope technique with 14 days between administrations of test meals labelled with $^{57}$Fe and $^{58}$Fe was used in this study. No information is available on the amount of unlabelled formula consumed by the infants and whether the volume varied between the two separate administrations. Iron absorption was found to be similar from both test meals; 5.8±7% vs 6.3±4% (formula vs formula with rice cereal).

#### *11.2.2.2.2 Complementary Foods*

Several studies have reported on iron bioavailability from different complementary foods.[8,10,17,36–38] For example, recent studies have reported on the enhancing effect of meat on non-heme iron absorption from a vegetable purée and the positive effect of ascorbic acid on erythrocyte incorporation of iron from complementary foods.[17,38] Dephytinization of an infant cereal with relatively low native content of phytic acid and ample quantities of added ascorbic acid did not increase iron bioavailability.[37] Iron was added as ferrous sulfate in this study, resulting in relatively high geometric mean iron absorption (8.5 to 8.7%). However, although infant formulas are usually fortified with ferrous sulfate, iron fortification of cereal products is more complicated due to unacceptable organoleptic changes during storage and food preparation after addition of water-soluble iron compounds. Iron bioavailability of iron compounds currently used, or proposed to be used, in iron fortification programs of complementary foods such as infant cereals has been evaluated in two recent studies.[8,10] Labelled compounds, similar to the commercial equivalents, were prepared for these studies and administered to healthy infants. The results show that iron bioavailability from infant cereals can be significantly improved by replacing ferric pyrophosphate with ferrous fumarate (geometric mean bioavailability 1.3% vs 4.1%).[10] In a study by Fox et al., iron bioavailability from iron glycine chelate was similar to that from ferrous sulfate when added to an inhibitory cereal meal or to a vegetable meal.[8] Thus, chelation of iron did not improve iron bioavailability in the presence of dietary inhibitors.

*11.2.2.2.3 Iron Supplements*

A vitamin-iron supplement, containing about 8 mg total iron/dose (labelled with $^{58}Fe$) and 26 mg ascorbic acid was given between feedings to 56-day-old infants on 3 consecutive days.[20] Erythrocyte incorporation of $^{58}Fe$ was significantly higher in breast-fed infants (7.8±2.4%) compared to formula-fed infants (4.4±3.9%), confirming the earlier data from this group regarding differences in red blood cell incorporation between infants fed human milk or formula.[19] The mean erythrocyte incorporation in breast-fed infants corresponded to a nutritionally significant amount, 0.62 mg. For each group, erythrocyte incorporation of iron was inversely correlated with plasma ferritin. However, although plasma ferritin values were higher in breast-fed infants than in infants fed formula, erythrocyte incorporation of iron was greater by the breast-fed infants.

Ten older infants, mean age 13 months, were given smaller doses of iron (5 mg, labelled with $^{57}Fe$ or $^{58}Fe$), followed by intake of cow's milk or apple juice.[39] Iron absorption was significantly greater from the iron supplement given with juice containing 42 mg added ascorbic acid; geometric mean absorption 11.8% (range 2.5 to 22.7%) than from the supplement given with milk (geometric mean absorption 4.6%, range 1.1 to 15.8%). An earlier report by the same investigators, using a similar study protocol, reported iron absorption in the range 1.5 to 2.0% (n = 3) from 11 mg $^{57}Fe$ and 2.4 to 9.7% (n = 4) from 5 mg iron labelled with $^{58}Fe$ in one-year-old infants.[40] However, the very limited number of children included in this study and the non-standardized dietary intake after administration of the iron supplements complicate the interpretation of the data.

### 11.2.2.3 Studies in Pregnant Women

Stable isotope techniques have been used in a few studies to monitor iron metabolism in pregnant women. As discussed earlier in this chapter, the methodology based on erythrocyte incorporation of iron stable isotopes 14 days after administration includes estimates of blood volume to calculate circulating iron. Thus, the expansion of plasma volume and red cell mass during pregnancy is a major methodological problem when using this technique in pregnant women, necessitating direct measurements of blood volume in study subjects.[41] In addition, erythrocyte incorporation rate can be assumed to be different in pregnant women as compared to non-pregnant subjects and to vary between early pregnancy and late pregnancy. Therefore, direct measurements of red cell incorporation after intravenous injection of an iron stable-isotope label are essential in studies during pregnancy. In particular, in longitudinal studies to evaluate changes in iron absorption during pregnancy, repeated measurements of blood volume and erythrocyte incorporation rate need to be included in the study protocol.

A cross-sectional study of 12 pregnant women (23 to 38 weeks of gestation) reported erythrocyte incorporation of 64.7±12.2% after intravenous administration of 0.7 mg iron (labelled with $^{58}Fe$). Iron absorption from an oral dose of iron (4 to 7 mg, labelled with $^{58}Fe$) was 50.7±21.3%.[41] Since $^{58}Fe$ was given twice, oral administration and intravenous injection of the stable isotope labels were separated by 14 days. Blood volume was measured by tagging red blood cells with $^{50}Cr$. In addition, analysis of cord blood samples was used to calculate the "transfer coefficient" of $^{58}Fe$, i.e., transfer of $^{58}Fe$ from maternal plasma to the infant at birth. The coefficient was 6.62± 3.69%. A cross-sectional study of iron absorption in pregnant Peruvian women (30 to 36 weeks gestation) was reported recently.[42] In this study, iron absorption was measured after oral intake of 10 mg $^{57}Fe$ in a drink (no added ascorbic acid) and intravenous injection of 0.6 mg $^{58}Fe$. Women consuming prenatal iron (or iron plus zinc) supplements were given an additional 50 mg iron after intake of the drink with added $^{57}Fe$. Erythrocyte incorporation of $^{57}Fe$ and $^{58}Fe$ was determined after 14 days. Women in the control group (no prenatal supplements, n = 17) incorporated significantly more $^{58}Fe$ into erythrocytes (91.5±28.0%) as compared with women taking supplements (76.4±13.1%, n = 28). No differences were observed in iron absorption; mean values were 9.9 to 13.0%. Blood volume was not measured in this study, but estimated based on 70 ml/kg. The magnitude of the error introduced in the calculations due to inaccurate blood volume determination is not known.

An alternative method, based on measurements of the "area under the curve" (AUC) after oral intake of $^{54}Fe$ (2.8 to 5.0 mg/dose) and intravenous injection of $^{57}Fe$ (200 μg), followed by repeated blood sampling over 6 to 10 hours and analysis of iron stable isotopes in serum, has been used in two studies. Longitudinal studies in nine women reported geometric mean iron absorption to increase significantly from 7.6% at 12 weeks of gestation to 21.1% (24 weeks), 36.3% (36 weeks), and 26.3% at 12 weeks postpartum.[43] A water solution containing 5.2 mg iron (5.0 mg $^{54}Fe$) was given after an overnight fast, followed by a light breakfast. When iron absorption was measured from a breakfast meal (including orange juice) labelled with 2.8 mg $^{54}Fe$, significant increases in iron absorption were observed in 12 women studied longitudinally. Geometric mean iron absorption increased from 7% (12 weeks of gestation) to 36% (24 weeks) and 66% (36 weeks). At 16 to 24 weeks postpartum, geometric mean iron absorption was 11%.[44]

## 11.3 Zinc

Besides iron, zinc is the trace element most frequently studied by stable-isotope techniques in infants. Most studies of zinc absorption to date have been based on fecal monitoring techniques in infants and in pregnant women.[3,15,21–23,25,26,46,47] These studies include measurements of zinc absorption from labelled human milk, preterm human milk, fortified preterm human milk and preterm formula, milk-based infant formula, infant cereals, and a vegetable-based weaning food.[15,21–23,25,45–47] In most infant studies, complete fecal collections over 72 hours, 3 to 4 days,[45] or 96 hours[21] were made. Extended collections of fecal material (21 days) were included in the study protocol to allow estimates of re-excretion of absorbed zinc in one study.[22]

Of special interest is the study by Serfass et al., where zinc absorption from intrinsically and extrinsically labelled infant formula was compared and the extrinsic labelling technique was validated for milk-based diets in infants.[23] Data reported by the same investigators demonstrated that infants could maintain zinc balance at low-zinc intake by increasing fractional zinc absorption and decreased excretion of endogenous zinc.[15] Zinc absorption from 940 ml labelled milk-based formula with "high" or "low" zinc concentration (6.58 or 1.47 mg/l) was measured in six healthy infants. Fractional $^{70}$Zn absorption increased significantly from 16.8±5.8% to 41.1±7.8% when formula with "high" zinc concentration was compared with "low" zinc concentration; endogenous zinc fecal excretion decreased significantly from 78±56 μg/kg/day to 34±16 μg/kg/day.

Zinc absorption from human milk, measured in nine infants 2- to 5-months-old, was reported to be relatively high, 54±7.5%.[26] In this study, six small aliquotes of labelled human milk (about 5 ml each; total dose $^{70}$Zn 153 to 213 μg) were administered orally via syringes before routine feeds over 24 hours. Endogenous fecal zinc, estimated in seven infants, was 0.31±0.15 mg/day (0.05±0.02 mg/kg/day). High fractional zinc absorption from a dose of $^{70}$Zn has been reported in 13 premature infants, range 48 to 79%.[46] Doses of $^{70}$Zn varied depending on feeding regimen (31.4 to 131 μg/kg) and were administered via nasogastric tube during a gavage feeding of preterm human milk, fortified preterm human milk or premature infant formula. Fractional zinc absorption did not appear to be related to postnatal age, postconceptual age, body weight, or diet in this study.

Very limited information about zinc absorption from complementary foods is available. A serving of an infant cereal based on white wheat and cow's milk, labelled with 370 μg $^{70}$Zn and fed to six healthy infants, resulted in zinc absorption of 33.9±16.4%.[22] Analysis of separate stools, collected during 21 days, confirmed that 72 hours is a sufficient time period for complete collections of non-absorbed stable isotopes in infants consuming semi-solid foods. Re-excretion of absorbed $^{70}$Zn (>68 to 92 hours to 21 days after intake) was very small, 0.44±0.38%. Dietary fiber did not significantly influence zinc absorption from infant cereals based on wheat and soy (8.0 vs 1.8% dietary fiber); 45.3±27.5% and 41.2±19.4% were absorbed by healthy infants.[25] Zinc absorption from a vegetable-based infant food, labelled with 1 mg $^{67}$Zn or $^{70}$Zn was 28.6±10.5% in 11 nine-month-old infants. Iron fortification of the test meal did not influence zinc absorption significantly.[45]

Fractional zinc absorption based on the double isotope method developed by Friel et al. in adults[48] has so far been reported in very-low-birth-weight infants, in five- to seven-month-old infants and from a longitudinal study during pregnancy and lactation.[2,5,24] The double isotope technique has the advantage of eliminating the need for fecal collections but necessitates intravenous administration of a zinc stable isotope. The acceptability of this methodology, in particular for studies in healthy infants, can therefore be expected to be limited. In addition, doses of zinc stable isotopes, particularly for intravenous administration, have not been optimized for studies in infants. Doses administered to infants would seem relatively high, 50 to 100 μg $^{70}Zn$/kg body weight[24] or 500 μg $^{67}Zn$[5] as compared to doses administered i.v. to adults (0.8 mg $^{70}Zn$ or 2.8 to 2.9 mg $^{68}Zn$)[48] and to pregnant or lactating women (0.8 mg $^{70}Zn$).[2]

The study by Friel et al. compared zinc absorption based on the double isotope method with fecal monitoring in 12 very-low-birth-weight preterm infants, resulting in similar fractional absorption: 0.22±0.09 and 0.25±0.07.[24] Endogenous fecal zinc excretion was 390±270 μg/kg/day. $^{68}Zn$ was administered with either human milk or formula, depending on the normal feeding regimen. Zinc absorption from labelled human milk was 49.5±18.5% in 10 five- to seven-month-old infants.[5] The study by Fung et al. provides new, important information on the dynamic changes in zinc metabolism during lactation.[2] When measured longitudinally in 13 women, a nearly twofold increase in zinc absorption was found during lactation (mean 25%), as compared with preconception data (mean 14.6%). Mean zinc absorption increased to 19% at 34 to 36 weeks of gestation, but this difference was not statistically significant as compared with preconception data.

In addition to information about zinc absorption, estimates of endogenous fecal zinc losses are of major importance to evaluate zinc homeostasis.[49,50] The use of intravenous administrations of zinc stable isotopes to estimate endogenous fecal zinc loss has been very limited in infants.[24,51] In most studies, calculations of endogenous zinc loss have been based on data obtained after oral administration of zinc stable isotopes.[15,26,47,49] The importance of endogenous zinc losses in infant nutrition has been demonstrated and discussed by Ziegler et al. and Krebs et al.[15,49] The data on increased fecal endogenous losses of zinc as a principal cause of zinc depletion in infants with cystic fibrosis are of special interest in clinical nutrition.[49]

More detailed data on zinc kinetics have been reported for infants and lactating women.[51,52] However, the need for repeated venipuncture, as well as complete collection of excreta, has limited these studies to only two preterm infants and five lactating women. The application of an alternative methodology to study zinc metabolism, based on stable-isotope dilution principles after intravenous administration of $^{67}Zn$ in free-living, lactating women, demonstrates an interesting possibility to implement stable isotopic techniques under field conditions.[52]

## 11.4 Zinc and Copper

Two studies of zinc and copper absorption, measured by stable-isotope technique in infants, have been published. In the study by Johnson and Canfield, stable-isotope labels of zinc and copper were administered simultaneously to breast-fed and formula-fed infants.[21] Infants were fed 30 ml human milk or formula, labelled with 2 mg $^{67}$Zn and 1 mg $^{65}$Cu, followed by ingestion of unlabelled milk or formula. The addition of stable isotopes changed the total content of both zinc and copper substantially in this study. Fecal material was collected for 96 hours. Fractional zinc absorption was statistically significantly higher in nine breast-fed infants (85.6±4.3%) as compared with seven formula-fed infants (65.5±12.1%). Percent copper absorption was very high from both test meals; mean absorption was 89.7% and 81.4% in breast-fed and formula-fed infants, respectively. Zinc and copper absorption from preterm formula (n = 33), preterm human milk (n = 7), fortified preterm human milk (n = 5), and term formula (n = 5) was measured in very-low-birth-weight infants.[47] Labelled feeds were prepared by the addition of $^{70}$Zn and $^{65}$Cu (40.1 to 120.2 µg $^{70}$Zn/kg body weight and 49.1 to 147.3 µg $^{65}$Cu/kg body weight) to human milk or formula. Feces were collected for 72 hours. $^{70}$Zn and $^{65}$Cu absorption was significantly higher from preterm human milk (mean 68.6% and 69.8%) than preterm formula (mean 31.6% and 39.6%) and term formula (mean 17.6% and 26.5%). Infants fed fortified preterm human milk absorbed an average of 48.4% ($^{70}$Zn) and 57.4% ($^{65}$Cu). Estimates of endogenous fecal zinc and copper losses were significantly lower in infants fed preterm human milk as compared with infants fed preterm formula.

The study in ten pregnant and five non-pregnant women, comparing zinc absorption from diets based on animal products with a plant-based diet, included measurements of copper absorption in parallel.[3,4] The diets were labelled with 10 mg zinc (enriched in $^{70}$Zn) and 3 mg $^{65}$Cu. Isotope doses were given in four meals, during one day. Feces were collected for 12 days. No differences in mean $^{70}$Zn absorption (about 25%) were found between diets or when pregnant and non-pregnant women were compared. Mean fractional absorption, and the amount of copper absorbed, from the plant-based diet were significantly higher in pregnant (n = 4) than in non-pregnant women (n = 5); 40.7% vs 33.8% and 1.03 vs 0.85 mg/day. Mean copper absorption from the animal protein diet was 41.2% and 42.2% in five non-pregnant and five pregnant women, respectively.

## 11.5 Selenium

Although selenium has been the focus of many nutritional studies, very few reports on selenium metabolism in infants or pregnant or lactating women based on stable-isotope technique have been published. Contrary to most other trace elements, considerable amounts of selenium are excreted in urine and retention is therefore a more useful measure of selenium metabolism than absorption. Ehrenkranz et al. evaluated the influence of increased selenium content in preterm infant formula on selenium absorption and retention of

$^{74}$Se-selenite in very-low-birth-weight infants.[53] Feces were collected for 72 hours and urine during 96 hours. Results from this study demonstrated that mean $^{74}$Se absorption was significantly different in the two groups; 91.2% vs 86.2% (13.4 vs 20.3 μg Se/L formula). However, mean fractional selenium retention was similar, 95.0 to 96.6%, in both study groups. Estimates of endogenous fecal losses resulted in significantly higher selenium loss in infants fed the selenium-supplemented formula; 1.14±0.76 μg/kg/day compared with 0.60±0.56 μg/kg/day. Recently, selenium absorption and retention from selenite and selenate were compared in healthy, formula-fed infants.[9] Equal quantities of infant formula (500 g) were labelled with 10 μg $^{74}$Se-selenite or 10 μg $^{76}$Se-selenate and administered in alternate feeds to nine infants. Excreta (urine and feces) were collected for 72 hours. Mean selenium absorption was significantly higher from selenate (97.3%) than selenite (73.4%), although selenium retention from the two selenium compounds was similar, 60.6±4.9% and 64.4±5.0%, respectively.

Two studies have reported on the use of stable-isotope technique to study selenium metabolism in pregnant and lactating women. Swanson et al. monitored fecal and urine excretion of $^{76}$Se for 12 days in ten pregnant women after intake of intrinsically labelled eggs.[54] Mean apparent absorption was about 80% in both pregnant and non-pregnant women. However, pregnant women showed a tendency toward renal conservation of $^{76}$Se; cumulative urinary losses of $^{76}$Se were statistically significantly less in women during late pregnancy, as compared with non-pregnant women. Utilization of inorganic (selenite) and organic (selenomethionine; Se-Met) selenium was investigated in lactating women, non-lactating women 2 to 3 months postpartum, and never-pregnant women.[6,7] Stable isotope labels (26.5 μg $^{74}$Se as Se-Met and 41.6 μg $^{76}$Se as selenite) were given without food and then feces and urine were collected for 2 weeks. Mean absorption of $^{74}$Se from Se-Met was 96 to 97% in all groups, significantly higher than the absorption of $^{76}$Se from selenite (mean values 34 to 47%). Mean apparent retention of $^{74}$Se (Se-Met) was about 82 to 94% (268 to 301 nmol) in all groups, whereas lactating women retained significantly more (mean 43%; 224 nmol) of the $^{76}$Se (selenite) compared to non-lactating and never-pregnant women (mean 26 and 30%; 130 and 167 nmol, respectively). At 24 hours postdosing, all groups had significantly more $^{74}$Se in plasma than $^{76}$Se; significantly more $^{74}$Se from Se-Met than $^{76}$Se from selenite appeared in human milk over 48 hours.

## 11.6 Chromium

A recent study reported on the distribution of chromium ($^{53}$Cr) in serum, urine, and human milk in lactating women after intake of high doses of chromium over 3 days (400 μg $^{53}$Cr/day).[55] Minimum absorption was estimated from urinary excretion of $^{53}$Cr to 0.41% on day 1 and the cumulative absorption through day 6 to 0.60% of the total dose. $^{53}Cr$ was not detected in human milk. Furthermore, no significant changes in chromium concentration in milk were observed during the 90-day study, suggesting that human milk chromium is independent of intake. Data from this study estimated chromium intake in breast-fed infants to be very low (0.13 μg/day), well below the lower end of the range of estimated safe and adequate daily dietary intakes (10 μg/day) for infants 0 to 6 months.[56]

## 11.7 Conclusion

It is best hoped that the use of stable-isotope techniques in studies of trace-element metabolism during early life, as well as in pregnant and lactating women, will increase with expanding applications. Studies using stable-isotope techniques have provided new, important information about iron metabolism and iron bioavailability from infant diets. However, the information is still limited for other trace elements. Well-designed studies, based on stable-isotope techniques, could provide much needed information on trace element metabolism in vulnerable segments of the population. In particular, use of stable-isotope techniques to monitor changes in trace-element metabolism in longitudinal studies, for example during pregnancy and lactation as well as during growth and development in infants, could generate important, new information.

## References


1. Widness, J.A. et al., Erythrocyte incorporation and absorption of $^{58}Fe$ in premature infants treated with erythropoietin, *Pediatr. Res.*, 41, 416, 1997.
2. Fung, E.B. et al., Zinc absorption in women during pregnancy and lactation: a longitudinal study, *Am. J. Clin. Nutr.*, 66, 80, 1997.
3. Swanson, C.A., Turnlund, J.R., and King, J.C., Effect of dietary zinc sources and pregnancy on zinc utilization in adult women fed controlled diets, *J. Nutr.*, 113, 2557, 1983.
4. Turnlund, J.R., Swanson, C.A., and King, J.C., Copper absorption and retention in pregnant women fed diets based on animal and plant proteins, *J. Nutr.*, 113, 2346, 1983.
5. Abrams, S.A., Wen, J., and Stuff, J.E., Absorption of calcium, zinc, and iron from breast milk by five- to seven-month old infants, *Pediatr. Res.*, 39, 384, 1996.
6. Mangels, A.R. et al., Selenium utilization during human lactation by use of stable-isotope tracers, *Am. J. Clin. Nutr.*, 52, 621, 1990.
7. Moser-Veillon, P.B. et al., Utilization of two different chemical forms of selenium during lactation using stable isotope tracers: an example of speciation in nutrition, *Analyst*, 117, 559, 1992.

8. Fox, T.E., Eagles, J., and Fairweather-Tait, S.J., Bioavailability of iron glycine as a fortificant in infant foods, *Am. J. Clin. Nutr.*, 67, 664, 1998.
9. Van Dael, P. et al., Selenite and selenate absorption and retention in infants, *J. Pediatr. Gastroenterol. Nutr.*, 28, 595 (A41), 1999.
10. Davidsson, L. et al., Iron bioavailability in infants from an infant cereal fortified with ferric pyrophosphate or ferrous fumarate, *Am. J. Clin. Nutr.*, 71, 1597, 2000.
11. Zlotkin, S.H. et al., Determination of iron absorption using erythrocyte iron incorporation of two stable isotopes of iron ($^{57}Fe$ and $^{58}Fe$) in very low birthweight premature infants, *J. Pediatr. Gastroenterol. Nutr.*, 21, 190, 1995.
12. McDonald, M.C., Abrams, S.A., and Schanler, R.J., Iron absorption and red blood cell incorporation in premature infants fed an iron-fortified infant formula, *Pediatr. Res.*, 44, 507, 1998.
13. Kastenmayer, P. et al., A double stable isotope technique for measuring iron absorption in infants, *Br. J. Nutr.*, 71, 411, 1994.
14. Davidsson, L. et al., Influence of lactoferrin on iron absorption from human milk in infants, *Pediatr. Res.*, 35, 117, 1994.
15. Ziegler, E.E. et al., Effect of low zinc intake on absorption and excretion of zinc by infants studied with $^{70}Zn$ as extrinsic tag, *J. Nutr.*, 119, 1647, 1989.
16. Moody, G.J., Schanler, R.J., and Abrams, S.A., Utilization of supplemental iron by premature infants fed fortified human milk, *Acta Paediatr.*, 88, 763, 1999.
17. Fairweather-Tait, S. et al., The bioavailability of iron in different weaning foods and the enhancing effect of a fruit drink containing ascorbic acid, *Pediatr. Res.*, 37, 389, 1995.
18. Davidsson, L. et al., Iron bioavailability in infants: The influence of phytic acid and ascorbic acid in infant formulas based on soy isolate, *Pediatr. Res.*, 36, 816, 1994.
19. Fomon, S.J., Ziegler, E.E., and Nelson, S.E., Erythrocyte incorporartion of ingested $^{58}Fe$ by 56-day-old breast-fed and formula-fed infants, *Pediatr. Res.*, 33, 573, 1993.
20. Fomon, S.J. et al., Erythrocyte incorporation of iron by 56-day old infants fed a $^{58}Fe$-labeled supplement, *Pediatr. Res.*, 38, 373, 1995.
21. Johnson, P.E., and Canfield, W.K., Stable zinc and copper absorption in free-living infants fed breast milk or formula, *J. Trace Elem. Exp. Med.*, 2, 285, 1989.
22. Davidsson, L. et al., Zinc and calcium apparent absorption from an infant cereal. A stable isotope study in healthy infants, *Br. J. Nutr.*, 75, 291, 1996.
23. Serfass, R.E. et al., Intrinsic and extrinsic stable isotopic zinc absorption by infants from formulas, *J. Nutr.*, 119, 1661, 1989.
24. Friel, J.K. et al., Zinc absorption in premature infants: comparison of two isotopic methods, *Am. J. Clin. Nutr.*, 63, 342, 1996.
25. Davidsson, L. et al., Dietary fiber in weaning cereals: a study of the effect on stool characteristics and absorption of energy, nitrogen and minerals in healthy infants, *J. Pediatr. Gastroenterol. Nutr.*, 22, 167, 1996.
26. Krebs, N.F. et al., Zinc homeostasis in breast-fed infants, *Pediatr. Res.*, 39, 661, 1996.
27. Janghorbani, N., Ting, B.T.G., and Fomon, S.J., Erythrocyte incorporation of ingested stable isotope of iron ($^{58}Fe$), *Am. J. Hematol.*, 21, 277, 1986.
28. Fomon, S.J. et al., Erythrocyte incorporation of ingested 58-iron by infants, *Pediatr. Res.*, 24, 20, 1988.
29. Ehrenkranz, R.A. et al., Iron absorption and incorporation into red blood cells by very low birth weight infants: studies with the stable isotope $^{58}Fe$, *J. Pediatr. Gastroenterol. Nutr.*, 15, 270, 1992.

30. Fomon, S.J. et al., Less than 80% of absorbed iron is promptly incorporated into erythrocytes of infants, *J. Nutr.*, 130, 45, 2000.
31. Fomon, S.J. et al., Time course of and effect of dietary iron level on iron incorporation into erythrocytes by infants, *J. Nutr.*, 130, 541, 2000.
32. Friel, J.K. et al., Elevated intakes of zinc in infant formulas do not interfere with iron absorption in premature infants, *J. Pediatr. Gastroenterol. Nutr.*, 27, 312, 1998.
33. Fairweather-Tait, S.J. et al., Lactoferrin and iron absorption in newborn infants, *Perdiatr. Res.*, 22, 651, 1987.
34. Fomon, S.J. et al., Erythrocyte incorporation of iron is similar in infants fed formulas fortified with 12 mg/L or 8 mg/L of iron, *J. Nutr.*, 127, 83, 1997.
35. Lifschitz, C.H., and Abrams, S.A., Addition of rice cereal to formula does not impair mineral bioavailability, *J. Pediatr. Gastroenterol. Nutr.*, 26, 175, 1998.
36. Fomon, S.J. et al., Iron absorption from infant foods, *Pediatr. Res.*, 26, 250, 1989.
37. Davidsson, L. et al., Iron bioavailability from infant cereals by infants: the effect of dephytinization, *Am. J. Clin. Nutr.*, 65, 916, 1997.
38. Engelmann, M.D.M. et al., The influence of meat on non-heme iron absorption in infants, *Pediatr. Res.*, 43, 768, 1998.
39. Abrams, S.A. et al., Absorption by 1-year old children of an iron supplement given with cow's milk or juice, *Pediatr. Res.*, 39, 171, 1996.
40. Abrams, S.A. et al., Application of magnetic sector thermal ionization mass spectrometry to studies of erythrocyte iron incorporation in small children, *Biol. Mass Spectrom.*, 23, 771, 1994.
41. Dyer, N.C. and Brill, A.B., Use of the stable tracers $^{58}$Fe and $^{50}$Cr for the study of iron utilization in pregnant women, in *Nuclear Activation in the Life Sciences*, The International Atomic Energy Agency (IAEA), Vienna, 1972, 469.
42. O'Brien, K.O. et al., Influence of prenatal iron and zinc supplements on supplemental iron absorption, red blood cell iron incorporation, and iron status in pregnant Peruvian women, *Am. J. Clin. Nutr.*, 69, 509, 1999.
43. Whittaker, P.G., Lind, T., and Williams, J.G., Iron absorption during normal human pregnancy: a study using stable isotopes, *Br. J. Nutr.*, 65, 457, 1991.
44. Barrett, J.F.R. et al., Absorption of non-haem iron from food during normal pregnancy, *Br. Med. J.*, 309, 79, 1994.
45. Fairweather-Tait, S.J., Wharf, G., and Fox, T.E., Zinc absorption in infants fed iron-fortified weaning foods, *Am. J. Clin. Nutr.*, 62, 785, 1995.
46. Ehrenkranz, R.A. et al., Determination with stable isotopes of the dietary bioavailability of zinc in premature infants, *Am. J. Clin. Nutr.*, 40, 72, 1984.
47. Ehrenkranz, R.A. et al., Zinc and copper nutritional studies in very low birth weight infants: comparison of stable isotopic extrinsic tag and chemical balance methods, *Pediatr. Res.*, 26, 298, 1989.
48. Friel, J.K. et al., The analysis of stable isotopes in urine to determine the fractional absorption of zinc, *Am. J. Clin. Nutr.*, 55, 473, 1992.
49. Krebs, N.F. et al., The use of stable isotope techniques to assess zinc metabolism, *J. Nutr. Biochem.*, 6, 293, 1995.
50. Hambidge, M.K., Krebs, N.F., and Miller, L., Evaluation of zinc metabolism with use of stable-isotope techniques: implications for the assessment of zinc status, *Am. J. Clin. Nutr.*, 68(suppl.), 410S, 1998.
51. Wastney, M.E. et al., Zinc kinetics in preterm infants: a compartmental model based on stable isotope data, *Am. J. Physiol.*, 40, R1452, 1996.
52. Jackson, M.J. et al., Stable isotope metabolic studies in zinc nutrition in slum-dwelling lactating women in the Amazon valley, *Br. J. Nutr.*, 59, 193, 1988.

53. Ehrenkranz, R.A. et al., Selenium absorption and retention by very-low-birth-weight infants: studies with the extrinsic stable isotope tag $^{74}$Se, *J. Pediatr. Gastroenterol. Nutr.*, 13, 125, 1991.
54. Swanson, C.A. et al., Quantitative and qualitative aspects of selenium utilization in pregnant and nonpregnant women: an application of stable isotope methodology, *Am. J. Clin. Nutr.*, 38, 169, 1983.
55. Mohamedshah, F.Y. et al., Distribution of a stable isotope of chromium ($^{53}$Cr) in serum, urine, and breast milk in lactating women, *Am. J. Clin. Nutr.*, 67, 1250, 1998.
56. National Research Council, *Recommended Dietary Allowances*, 10$^{th}$ ed., National Academy Press, Washington, D.C., 1989.

# 12

# *Stable-isotope Studies in the Elderly*

**Catherine I.A. Jack, Nicola M. Lowe, and Malcolm J. Jackson**

**CONTENTS**

12.1 Introduction ........ 187
12.2 Practicalities of Working with Elderly Subjects ........ 188
12.3 Ethical Considerations ........ 188
12.4 Examples of Stable-isotope Studies in the Elderly ........ 189
  12.4.1 Zinc Homeostasis in the Elderly ........ 189
  12.4.2 Copper Homeostasis in the Elderly ........ 189
12.5 Selenium Status of the Elderly ........ 190
12.6 Conclusion ........ 190
Acknowledgments ........ 190
References ........ 191

## 12.1 Introduction

Nutrition is recognized as an important factor in age-related diseases such as cancers, cardiovascular disease, osteoporosis, and cataract. Elderly patients appear to be at increased risk of malnutrition because of multiple factors, such as restriction of activity, decrease in autonomy, multiple medications, and decreases in appetite.[1] Few studies of trace elements in the elderly have been undertaken and, of these, even fewer have reached definitive conclusions concerning whether a specific nutritional problem occurs in this group. Isotopic techniques offer the possibility of clarifying this situation, although so far only a handful of studies have been undertaken.

All studies in the elderly involve taking into account a number of complications that are specific to this age group. This chapter will discuss these complications prior to a brief review of some of the work which has been published in this area.

0-8493-8730-2/00/$0.00+$.50

## 12.2 Practicalities of Working with Elderly Subjects

Because of the heterogeneity of the population described as "elderly" or "aged," individuals recruited to the study must be clearly defined. Typical criteria might be "healthy elderly people over the age of 75," or be defined more specifically as "elderly people over the age of 75 years with pressure sores grade 2 or above." Members of any "healthy" group in this age range, to be included, must all undergo a detailed medical history, including previous operations and medical illnesses. Very few, if any, in this age range will have no evidence of previous disease. A drug, substance, and allergy history should also be ascertained to determine if the subject is on any substance likely to interfere with the study or is likely to have an anaphylactic or allergic reaction to the micronutrient or carrier medium. A thorough examination is also required, and baseline blood tests are done to ensure subjects do not have kidney or liver impairment and are not anemic or have any unknown endocrine problems that could affect how their bodies handle micro-nutrients and their excretion, etc.

Other problems may occur in isotopic trace-element studies in elderly people that include rigorous protocols with multiple sampling over several days or even weeks. This can lead to practical problems if the older person is a healthy volunteer and has to attend hospital but does not have access to personal transport. Alternatively, samples can be taken in the individual's own home or institution, but there must be systems in place to allow quick separation of blood, etc. Older people may also be put off a clinical study by the type of samples required, such as urine or feces. These can also cause collection problems in those with visual or memory problems, and can be even more difficult in incontinent individuals.

In studies of disease states in the elderly, multiple pathology can also limit recruitment of elderly subjects. Thus, for example, it is very difficult to find elderly diabetics who do not have other underlying pathologies, such as cardiac impairment, hypertension, and renal impairment. Additionally, polypharmacy can interfere with trace element studies, particularly as drugs such as diuretics may interfere with excretion.

## 12.3 Ethical Considerations

Studies in the elderly can present specific ethical problems, since researchers, health professionals, and ethics committees that approve research protocols must try to ensure that any agreement to participate in research represents the free choice of each individual. Frail, confused elderly patients may wish to please the researcher or potentially not understand what is being proposed, so it is important to ensure that informed consent is correctly obtained. To this end, it is vitally important to provide adequate advice and information in understandable language to allow older people to make a valid and independent decision. In practice, it is necessary that older individuals are given appropriate time to consider the study and to be able to discuss it with both the researchers and those who care for them. It is also important that the elderly person is aware that he is under no obligation to agree to participate, his treatment or care will not be affected if he decides not to get involved, and that, if he does agree to participate, he can withdraw at any time and does not have to explain why he is withdrawing.

## 12.4 Examples of Stable-isotope Studies in the Elderly

### 12.4.1 Zinc Homeostasis in the Elderly

A number of studies have examined the zinc status of the elderly with conflicting results. Sandstead et al. reviewed the literature in 1982 and concluded that the elderly have a reduced zinc intake and that a proportion of elderly subjects are zinc-deficient, based on both dietary and laboratory data.[2] This appears to be particularly true in the hospitalized elderly where the proportion of subjects deficient in zinc may be as high as 67%.[3–6]

Isotopic studies of zinc in the elderly have examined the ability of young and old subjects to adapt to maintain homeostasis at different dietary levels of zinc. Turnlund et al., found that healthy elderly subjects had a reduced absorption of zinc compared with young when consuming diets of equivalent zinc content; this was confirmed by August et al.[7,8] Couzy et al. did not find similar changes in a group of healthy elderly Swiss subjects.[9] In our recent work we have confirmed the findings of Turnlund et al. and August et al. indicating that healthy elderly subjects have a reduced absorption of zinc compared with young when consuming diets of equivalent zinc content.[10] A rise in zinc intake (by 10 mg/day) also caused no change in the size of the rapidly exchangeable zinc pools in either group, but young subjects showed an adaptive fall in their fractional rate of zinc absorption whereas this was not seen in the elderly. The overall conclusion from this work was that aging impairs the ability of the gastrointestinal tract to respond to changes in dietary zinc intake.

### 12.4.2 Copper Homeostasis in the Elderly

The situation with copper homeostasis in the elderly is similar to that of zinc. There is some debate over whether currently available indicators of copper status are adequate for use in situations where marginal deficiency may occur.[11] The elderly may represent such a situation and conflicting data have been published concerning copper status in this group.

Stable isotopes have been used only peripherally to address this problem with studies of gastrointestinal copper absorption in elderly subjects.[8,12] No significant differences between young and elderly subjects were seen, although there was some variation between the two studies.[8] Further work appears necessary in this area.

## 12.5 Selenium Status of the Elderly

As part of more general concerns about potential deficits in the antioxidant nutrients in elderly subjects, Ducros and co-workers studied the sizes of two exchangeable body selenium pools in two groups of elderly women in comparison with young subjects.[13] Although dietary selenium intakes were equivalent between the different groups, the institutionalized elderly had lower plasma selenium levels and the total selenite pool (sum of the two pools measured) was reduced in comparison with "free-living" elderly and young control subjects. The authors concluded that the selenium status of elderly women is more related to lifestyle than age *per se*.

## 12.6 Conclusion

These brief examples illustrate the applicability of isotopes to studies in elderly subjects and the relevant data which can be obtained. The inherent safety of stable isotopes makes them particularly applicable to studies in the elderly in whom excretory routes may be compromised and this, combined with demographic changes leading to an increasingly aged population, is likely to lead to an expansion of activity in this area.

## Acknowledgments

The authors would like to acknowledge the support of the Wellcome Trust and the NHS NW Regional R&D Fund for financial support of their work.

## References

1. Vellas, B.J., Sachet, P., and Baumgartner, R.J., *Nutritional Intervention and the Elderly. Facts and Research in Gerontology.* Supplement: Nutrition, Serdi Publishing Co., New York, 1995, 5.
2. Sandstead, H.H., Henriksen, L.K., Greger, J.L., Prasad, A.S., and Good, R.A., Zinc nutrition in the elderly in relation to taste acuity, immune response and wound healing, *Am. J. Clin. Nutr.*, 36, 1046–1059, 1982.
3. Senapati, A., Jenner, G., and Thompson, R.P.H., Zinc in the elderly, *Quart. J. Med.*, 261, 81–87, 1989.
4. Stafford, W., Smith, R.G., Lewis, S.J., Henery, E., Stephen, P.J., Rafferty, J., Simpson, G.K., Bell, P.C., and O'Rorke, K., A study of zinc status of healthy institutionalised patients, *Age and Ageing*, 17, 42–48, 1988.
5. Thomas, A.J., Bunker, V.W., Hinks, L.J., Soda, N., Mullee, M.A., and Clayton, B.E., Energy, protein, zinc and copper status of twenty-one elderly inpatients: analysed dietary intake and biochemical indices, *Br. J. Nutr.*, 59, 181–191, 1988.
6. Paterson, P.G., Christensen, D.A., and Robertson, D., Zinc levels in hospitalized elderly, *J. Am. Diet. Assoc.*, 85, 186–191, 1985.
7. Turnlund, J.R., Durkin, N., Costa, F., and Margen, S., Stable isotope studies of zinc absorption and retention in young and elderly men, *J. Nutr.*, 116, 1239–1247, 1986.
8. August, D., Janghorbani, M., and Young, V., Determination of zinc and copper absorption at three dietary Zn-Cu ratios in young and elderly subjects, *Am J. Clin. Nutr.*, 50, 1457–1463, 1989.
9. Couzy, F., Kastenmayer, P., Mansourian, R., Guinchard, S., Munoz-Box, R., and Dirren, H., Zinc absorption in healthy elderly humans and the effect of diet, *Am. J. Clin. Nutr.*, 58, 690–694, 1993.
10. Ali, S., Lowe, N.M., Jack, C.I.A., Reid, M.D., Beattie, J.H., King, J.C., and Jackson, M.J., Zinc absorption in the healthy elderly, *Proc. Nutr. Soc.*, 57, 69A, 1998.
11. Linder, M.C., Copper, in *Present Knowledge in Nutrtion*, Ziegler, E.E. and Filer, L.J., Eds., ILSI, Washington 1996, 307–319.
12. Turnlund, J.R., Michel, M.C., Keyes, W.R., Schutz, Y., and Morgan, S., Copper absorption in elderly men determined by using stable $^{65}$Cu, *Am. J. Clin. Nutr.*, 36, 587–591, 1982.
13. Ducros, V., Faure, P., Ferry, M., Couzy, F., Biajoux, I., and Favier, A., The size of the exchangeable pools of selenium in elderly women and their relation to institutionalization, *Br. J. Nutr.*, 78, 379–396, 1997.

# 13

# *Applications of Trace-element Studies in Developing Countries: Practical and Technical Aspects*

**R.S. Gibson and C. Hotz**

**CONTENTS**

13.1 Introduction ........ 194
13.2 Applications of Isotope Studies in Developing Countries ........ 195
13.2.1 Supplementation ........ 195
13.2.2 Fortification ........ 197
13.2.3 Dietary Strategies ........ 198
13.3 Practical Aspects of Implementing Isotope Studies in Developing Countries ........ 199
13.3.1 Securing Support within the Country at the National and Community Level ........ 199
13.3.2 Selecting the Study Design ........ 200
13.3.3 Assessing the Nutritional and Health Status of the Study Participants ........ 201
13.3.4 Assessing Levels of Trace Elements and Absorption Modifiers in the Habitual Diets of Study Participants ........ 203
13.3.4.1 Assessing Food Intakes ........ 203
13.3.4.2 Compiling a Local Food Composition Table for Use in a Developing Country ........ 204
13.3.4.3 Assessing Intakes of Trace Elements and Absorption Modifiers in Habitual Diets ........ 204
13.3.4.4 Assessing Nutrient Intakes during the Metabolic Study ........ 205
13.4 Technical Aspects of Implementing Isotope Studies in Developing Countries ........ 206
13.4.1 Considerations When Selecting the Isotopic Technique ........ 207
13.4.1.1 Fecal Monitoring ........ 207
13.4.1.2 Urinary Monitoring ........ 208
13.4.1.3 Tissue Retention ........ 209
13.4.1.4 Plasma Tolerance Curves and Plasma Deconvolution ........ 209

0-8493-8730-2/00/$0.00+$.50

13.4.2 Collecting, Preparing, and Processing the Metabolic Samples for Analysis of Native Trace Elements and Isotopic Enrichment ..210
13.4.2.1 Fecal Samples....210
13.4.2.2 Urine Samples....211
13.4.2.3 Blood Samples ....211
13.5 Conclusion ....212
References....212

## 13.1 Introduction

The study of trace-element deficiencies in developing countries, and the evaluation of intervention strategies for their prevention, provide unique opportunities for the application of stable-isotope techniques. Such deficiencies arise from inadequate intakes, impaired absorption and/or utilization, excessive losses, or a combination of these factors. The deficiencies are exacerbated during times of greater physiological need such as infancy, pregnancy, lactation, and catch-up growth following illness. In population groups in developing countries, dietary diversity, food consumption patterns, physiological condition, and health status all differ markedly from those in developed countries. All these factors are known to modulate mineral bioavailability to varying degrees, making it essential to study mineral metabolism in selected high-risk population groups in developing countries to obtain data relevant to their needs.

In 1990, the World Health Organization (WHO), United Nations Children's Fund (UNICEF), and the World Summit for Children endorsed the elimination of micronutrient malnutrition in developing countries by the year 2000, specifically deficiencies of vitamin A, and two trace elements — iodine and iron. In the Third United Nations Report on the World Nutrition Situation, a third trace element, zinc, was added to this list.[1] Isotope techniques are currently available to study the absorption and metabolism of vitamin A, iron, zinc, and iodine, as well as selenium and calcium, two additional inorganic nutrients often identified as "high risk" in certain regions.

Nutrition intervention strategies proposed to eliminate micronutrient malnutrition include: supplementation, fortification, and dietary diversification/modification. Isotopic methods can play a vital role in optimizing the impact of all these strategies on the micronutrient status of population groups, as well as in quantifying and monitoring their efficacy, and, ultimately, their effectiveness at the program level. They can also be used to establish trace-element requirements in population groups in developing countries whose habitual diets are frequently plant-based, and thus contain high levels of antinutrients (e.g., phytic acid and polyphenols), known to interfere with the bioavailability of certain trace elements, notably zinc, nonheme iron, and possibly copper and manganese.

This review highlights some of the numerous nutritional applications of isotopes in developing countries and emphasizes both the practical and technical aspects of their use.

## 13.2 Applications of Isotope Studies in Developing Countries

The nutritional adequacy of dietary trace elements depends on their amount and bioavailability in the diet. Bioavailability can be defined as the proportion of the total trace element in a food or diet that is absorbed and utilized for normal bodily functions. Two major factors influence the bioavailability of trace elements: physiological and dietary factors. Physiological factors include the trace element nutriture of the individual, his or her developmental and health status, as well as the existence of certain adaptive mechanisms. The latter may be especially evident in populations in developing countries consuming predominately plant-based diets. The dietary factors include the physicochemical properties of the trace elements in the food or diet (e.g., pH, solubility, charge density, state of oxidation), the presence of certain dietary modifiers which may form complexes (e.g., dietary fiber) or chelates (phytic acid) with the trace elements, and the existence of competitive antagonism between ions (e.g., Cu-Zn; Cu-Fe; Fe-Zn; Mn-Fe; Cd-Zn) during digestion and absorption. The trace-element content of most human diets in developing countries is generally not high enough to induce such competitive interactions, but they could become important if staple foods or diets are fortified with trace elements. Isotope techniques are valuable in studying the effects of these physiological and dietary factors on the bioavailability of inorganic nutrients in developing countries.

Three nutrition intervention strategies can be used to combat deficiencies of micronutrients (including the trace elements iodine, iron, zinc, and selenium): (1) supplementation, (2) fortification, and (3) dietary diversification/modification. Applications of isotope techniques for each of these strategies are discussed below.

### 13.2.1 Supplementation

Supplementation has been extensively used to combat deficiencies of iron. This strategy is especially suitable for population groups whose requirements for iron and other trace elements cannot be met from habitual dietary sources and whose trace-element status must be improved over a relatively short time frame (e.g., pregnant women, low-birth-weight infants, and/or malnourished infants).

Isotopic methods can be used to study the relative efficiency of absorption of different doses and forms of trace-element supplements consumed alone, or with a meal or whole diet rich in absorption inhibitors such as phytic acid or polyphenols. The bioavailability of a wide range of different iron supplements has been compared by measuring the incorporation of isotopically labelled oral iron into erythrocytes. In contrast, isotopic studies on the bioavailability of different forms of supplemental zinc, when consumed either alone or in the presence of plant-based diets, are very limited. Radioisotope studies have shown that absorption of zinc from supplements is greater from higher doses when consumed in the absence of food as aqueous solutions, but when the supplements are consumed with a meal, absorption decreases at higher intakes. Furthermore, the relationship between zinc content and absorbed zinc indicates a saturation of absorption at an intake of 70 to 80 μmol/meal resulting in 18 to 20 μmol Zn absorbed.[2] These results emphasize that, for optimal absorption, supplemental zinc taken with meals should be distributed over meals throughout the day, rather than being administered only at one meal. By contrast, human studies for manganese suggest that neither the mode of administration nor the level of manganese in humans appears to impact percentage absorption, although in rats absorption reportedly decreases as dietary manganese increases.[3,4]

To date, most trace-element supplements have been given singly rather than as a component of multi-micronutrient supplements, with the exception of prenatal iron and folate supplements. This is unfortunate because in habitual diets in developing countries, the content and bioavailability of several trace elements, notably iron, zinc, iodine, and selenium, may be poor. Currently, a multi-micronutrient supplement is being formulated by UNICEF for use by pregnant women in developing countries. Care must be taken when formulating such multi-micronutrient supplements to ensure that the chemical form of the micronutrients is readily absorbed, and levels proposed do not induce antagonistic trace-element interactions (e.g., Cu-Zn; Cu-Fe; Fe-Zn; Mn-Fe) and thus interfere with the utilization of trace elements in the supplements per se, and/or with the utilization of elements intrinsic to the food or the meal. More data are required to establish if such antagonistic interactions depend on whether multi-micronutrient supplements are given in the fasted or fed state. For example, it is known that excess iron can affect zinc uptake when iron and zinc are given together in a water solution and in a fasting state, but not when given in the presence of *dietary* ligands in a food or meal.[5,6] By contrast, high levels of zinc supplements have no effect on iron absorption measured by radioisotopic methods in adults when they are given either in solution or as part of a meal, even when a 300:1 molar excess of zinc-to-iron is used.[7] To our knowledge, comparable data on interactions between iron and copper, or iron and manganese, given as supplements in a water solution, or with a meal, are not available.

Future isotope studies should include investigations of the bioavailability of graded doses of trace-element supplements given either singly, or in combinations, in the fasting state, or with meals representative of those consumed in developing countries. Furthermore, because of the potential for antagonistic interactions, the intrinsic levels of other major and trace minerals in the habitual diets of the study population must also be taken into account.

### 13.2.2 Fortification

Fortification with multiple micronutrients may be a cost-effective and sustainable method for improving the trace-element status at a national level in countries where trace-element deficiencies are endemic. Alternatively, fortification can be targeted in specific regions and/or for certain high-risk groups (e.g., complementary foods for infants) within a country. Successful fortification depends on the existence of a food vehicle that is centrally processed, temperature-stable, technologically and economically fortifiable, and undergoes no changes in taste, texture, and appearance during storage. As well, intakes of the food product at a relatively low level of consumption must be sufficient to provide adequate intakes of the trace elements to the population most at risk of deficiency, whereas, at higher consumption levels, there should be no risk of toxicity. Information on methods of storage, food processing, and preparation of the potential food vehicle must also be available to assess any potential losses of the fortificant.[8]

In developed countries, the level of the fortificants added to cereals is based generally on restoration levels (i.e., adding enough to the refined flour to restore the level to that of the unrefined cereal). In developing countries, however, higher levels than those normally present in unrefined cereals are necessary. If the potential food vehicle and/or indigenous meals contain potent inhibitors of trace-element absorption (e.g., phytate), the added trace elements, like the intrinsic trace elements, may be poorly absorbed, and hence may have limited impact on the trace-element status of the consumer. In an effort to counteract this problem, *protected* fortificants that prevent trace elements from binding to inhibitors such as phytic acid should be used. To date, only a protected iron compound has been developed — iron sodium ethylene-diamine-tetra-acetate (FeNaEDTA). Isotope studies have shown that FeNaEDTA may even enhance the absorption of intrinsic inorganic iron and zinc from meals containing phytic acid.[9,10] Unfortunately, the effect of trace-mineral absorption enhancers (e.g., ascorbic acid on non-heme iron) may be blunted in the presence of EDTA-containing compounds. As a result, the use of sodium iron EDTA is not recommended in diets that contain an abundance of trace-element absorption enhancers. More research is required utilizing isotope techniques to establish the extent to which such EDTA-containing compounds impact trace-element absorption modifiers and their possible influence and physiological impact on absorption of potentially toxic metals (Pb, Hg, Al, Mn).[11] Isotope studies are also required to identify bioavailable protected zinc compounds. To date, results of one radioactive isotope study on dogs suggest that use of an amino acid chelate of zinc may act as a protected fortificant.[12]

Isotopic methods thus have an important role in establishing both the appropriate chemical form and the levels of the fortificant to be used, identifying a suitable food vehicle for the fortificant(s), and establishing the impact on absorption of dietary components both within the food vehicle and intrinsic to the various indigenous meals.

### 13.2.3 Dietary Strategies

Dietary strategies to combat trace element deficiencies involve methods to enhance access to and utilization of foods with a high content and/or bioavailability of micronutrients in household diets. The strategies, although long-term, are more sustainable, economically feasible, and culturally acceptable than supplementation and fortification. They can also be used to alleviate several micronutrient deficiencies simultaneously without the risk of antagonistic interactions. Dietary strategies can focus on increasing the content of trace elements in staple foods and/or diets. Alternatively, the focus can be on improving the absorption of trace elements by altering the level of absorption modifiers in plant-based staples and/or household diets. Isotope studies can play an important role in quantifying trace-element bioavailability in plant-based staples or household diets, and in quantifying the potential for dietary modification to improve trace-element status.

Examples of strategies to improve micronutrient intake and bioavailability include: (1) enhancing the mineral and trace element content of plant-based staples, such as maize or rice, through use of soil fertilizers, foliar applications, plant-breeding, or genetic engineering; (2) increasing the consumption of flesh foods; (3) improving the bioavailability through the reduction of inhibitors such as phytic acid and polyphenols by breeding low-phytate varieties of cereal crops (maize), enzymatic hydrolysis of phytic acid in cereals via germination, fermentation or soaking, and use of commercial phytase enzymes; (4) diffusing into water soluble phytate and polyphenols in cereals and legumes by soaking; and (5) improving the bioavailability of minerals through increased consumption of absorption enhancers such as animal muscle proteins (for non-heme iron and zinc), organic acids such as ascorbic acid (for non-heme iron), and citric, malic, lactic, and tartaric acids (for non-heme iron and zinc).[13–17]

Human studies, using isotope techniques, of the bioavailability of micronutrients in staple foods or whole diets modified to reduce their phytate content are limited. Sandberg et al., using radioactive isotopes, recently reported increased iron absorption (i.e., 26.1 vs 14.3%) in Swedish individuals fed single meals containing white wheat rolls supplemented with phytase-deactivated wheat bran but with added microbial phytase from *A. niger* compared to those receiving the same meal but with no added microbial phytase.[18] Mendoza et al. assessed the effect of genetically modified, low-phytic acid maize on absorption of iron from tortillas using the extrinsic labelling technique.[19] Iron absorption was determined as the amount of radio-iron incorporated into red blood cells of 14 non-anemic U.S. men two weeks after being fed tortillas prepared with low-phytic acid "flint" maize (LPM) or its parent, wild-type strain using a reference dose of ferrous ascorbate. Iron absorption was reported to be 49% greater from the tortillas prepared from LPM (8.2% of intake) compared to those from the wild-type-strain (5.5% of intake) ($p<0.001$), after adjusting results to 40% absorption of ferrous ascorbate.

These two studies highlight the potential usefulness of methods based on reducing the dietary phytate content for improving iron and zinc status in populations in developing countries that consume predominately cereal-based diets.

## 13.3 Practical Aspects of Implementing Isotope Studies in Developing Countries

Before planning any isotope studies in developing countries, consideration must be given to securing support for the study at national, regional, and community levels. The study design selected must also take into account any practical difficulties that may arise when working in urban or rural communities in the developing country. Potential problems to consider are access to running water, a reliable electricity supply, adequate refrigeration and sanitation facilities, as well as cultural or religious barriers associated with the collection of biological samples or fluids and the intravenous administration of isotopes.

### 13.3.1 Securing Support within the Country at the National and Community Level

The first step in undertaking isotope studies in developing countries is for the principal investigators to consult the senior nutritionists and health professionals from government agencies such as the Ministries of Health, Agriculture and Community Services, institutions such as universities, colleges, and possibly non-government organizations (e.g., UNICEF, Save the Children Fund) in the country. Once support for the project at this level has been granted, then ethical approval from the appropriate Human Ethics Committee of the country must be obtained, as well as from the collaborating institution or agency. In countries where a Human Ethics Committee does not exist, approval must be sought from the advisory/technical committee of the appropriate ministry. If community-based, rather than clinic-based, studies are being conducted, approval for the project must also be secured at the regional, district, and community level. This can be sought by the study coordinator, preferably a person with previous experience in nutrition studies in developing countries, who is known to and has the cooperation of appropriate government agencies within the country. In long-term studies, it is often helpful to set up a consultative committee of district health, nutrition, and/or agricultural officials, with whom the study coordinator can have regular meetings. The committee can provide advice about how best to work with the community where the study will be held and can report back to their Government Departments on the project activities. Local community leaders (religious, political, and cultural) should be informed of the purpose of the study and its relevance to the community, and their approval obtained.

Liaison with the community must be maintained throughout the entire study so that the community is sensitized to the study, aware of its importance, informed of its progress, understands where the information collected is likely to be used, and has an opportunity to raise any questions or concerns. At the end of the project some way of providing feedback to the participants and the community must be established and implemented and a final report of the project submitted to the appropriate ministry or collaborating institution.

### 13.3.2 Selecting the Study Design

There are two types of experimental designs that are used for isotope studies: within-group, time-series designs and between-group designs. The former can be used to compare the fractional absorption of test meals or diets in the *same* group of subjects, so that each subject serves as his/her own control. The choice of the design generally depends on the time and resources available for the study, the study group, and the level of compliance/attrition expected. The advantages and disadvantages of each are considered below.

Crossover, within-group, time-series designs are recommended when comparing the treatments to avoid the impact of any time-dependent, confounding variables on the study results. With this design, half of the study participants will be randomly assigned to start with the first labelled test meals or whole diets and then will receive the second treatment later while the other participants do the opposite. In this way the impact of any confounding variables on the comparisons of absorption between different test meals or diets is eliminated. Factors to consider include age, sex, trace-element status, previous dietary history, gastrointestinal transit time, and physiological factors such as overall nutritional status (e.g., degree of wasting), presence of infections (e.g., malaria, HIV), and the health and integrity of the gastrointestinal tract. The latter may be compromised by the presence of parasites, certain nutrient deficiencies (e.g., riboflavin, zinc), diarrheal infections, and, possibly, genetic factors. All these considerations are especially critical when the absorption and/or metabolism of the trace mineral under study is known to be influenced by these factors (e.g., iron and zinc). An additional advantage of this design is that a smaller number of subjects is required, which is of particular relevance in view of the high cost of stable isotopes, and the subsequent costs and time for the analysis. Note, however, that when using such a design with a single isotope to minimize the possibility of a carryover effect, a time lapse of 14 days is required before the second labelled test meal or whole diet can be fed. Such a time lapse means that intra-individual variation is less controlled, and the likelihood of attrition is increased.

Another option which avoids these problems is to use a dual-isotope method. In this method, absorption of two test meals is compared by labelling each meal with a different radioactive or stable isotope. Such a technique enables comparisons between different test meals to be made on consecutive days within the same subjects, allowing the study to be completed over a much shorter time frame than the single-isotope method, thus enhancing compliance. For the dual-isotope method, test meals are given after an overnight fast, with no other foods or fluids for four hours, after which a standardized diet is given to all subjects for the remainder of each of the two days. After the two days of feeding, an unrestricted diet can be given for the rest of the study period. When blood samples are required, they must also be taken on the days when the test meals are fed.

In circumstances where it is not feasible to use a within-group, time-series design, a between-group design is used whereby the outcomes are measured in separate groups of subjects, each group receiving different test meals or whole diets. To overcome the problem of baseline differences in the trace-element status between subjects, an adjustment can be made based on their initial serum trace element concentrations (e.g., ferritin; plasma zinc) by analysis of co-variance (ANCOVA), where initial serum concentrations are used as a co-variant.[20] Another approach uses measured absorption from a reference dose to correct for measured absorption from the test meal, thus accounting for any differences in the initial iron status of the participants. A randomized complete block design may also be used to account for variations in independent variables such as anthropometric indices (e.g., body mass index, or weight-for-height Z-scores) and age, particularly when studying infants and young children.

While the many potential confounding factors on mineral absorption and metabolism may not be unique to subjects in developing countries, they may be more prevalent and more difficult to avoid. Therefore, the inclusion and exclusion criteria for the subjects should be carefully selected, based on the specific study objectives. For example, although iron status is known to affect the bioavailability of iron, it may not be appropriate to include only subjects with a normal iron status, as iron depletion may be inherent in the target population. Consequently, the extent to which these factors may affect the interpretation of the results must be carefully considered, as well as how the results can be extrapolated from the sample to the overall target population.

### 13.3.3 Assessing the Nutritional and Health Status of the Study Participants

Assessment of the nutritional and health status of the participants in the isotope study must be carried out at baseline as well as during the isotope study. The baseline data are essential for selecting subjects who meet the inclusion criteria of the study, for defining the baseline nutritional status of the study group(s), and for assisting in the interpretation of the isotopic results, as discussed above. Baseline data on nutritional status are especially critical for those isotope bioavailability studies in which absorption (e.g., iron and zinc) is affected by the initial trace-element status of the subjects.

Selection of the most appropriate laboratory indices for assessing trace-element status depends on the trace element under study. Several laboratory tests are available; a summary of recommended biochemical tests for iron, selenium, and zinc is given by Gibson.[21] When between-group study designs are employed, the baseline laboratory results of trace-element status can be used to match the participants prior to randomly assigning each member of the pair to a diet group. In this way, no significant differences in the baseline trace-element status are apparent between each group at the beginning of the study. However, with this design, biochemical analysis must either be made within country (e.g., hemoglobin for iron), or time allowed for the samples to be sent out of the country for analysis and results sent back to the study site (e.g., plasma and zinc). Alternatively, differences in the initial trace-element status can be adjusted using trace-element concentrations as a co-variant, as noted earlier. Appropriate precautions must be undertaken to avoid adventitious contamination during the collection, transfer, storage, handling, and analyses of any biopsy material taken for assessing trace-element status; this is discussed in more detail below.

Many factors may impact laboratory tests of trace-element status and confound the interpretation of the results, apart from depleted body stores of a trace element. For example, for the trace elements iron, zinc, and copper, circulating levels in the blood are altered by concurrent infection or inflammatory stress, when levels reflect a re-distribution in body compartments rather than deficiency or excess. In such cases, study subjects with concurrent infection must be identified by the assay of acute phase proteins in serum such as C-reactive protein (for chronic infection) or alpha-1-glycoprotein (for acute infection). More specific tests may also be used, such as those to identify intestinal parasites, malaria, or HIV.

Generally, a combination of laboratory tests is used, rather than a single test for each trace element; several concordant abnormal values are more reliable than a single aberrant value in diagnosing trace-element status. Recently, tests based on measurements of functional impairment (e.g., growth, body composition, cognitive function, immune competence, work capacity, morbidity) have become increasingly used as additional indices of iron and/or zinc status. Such tests have greater biological significance than the static biochemical tests because they measure the extent of the functional consequences of a trace-element deficiency. Hence, they are of particular interest in comprehensive community studies designed to assess the long-term impact of improving bioavailability of trace elements on subsequent growth, development, morbidity, and mortality in vulnerable groups in developing countries (i.e., infants, children, pregnant and lactating women).

In such community studies, anthropometric indices of nutritional status such as weight-for-length, height-for-age, and sometimes, weight-for-age and body-mass index, expressed as Z scores, are also used for matching participants. In this way, any potential baseline confounding variables are balanced across the treatment groups. The Z-scores are based on the United States National Center for Health Statistics (NCHS) reference growth data and can be calculated using a computer program[22] produced by Dean et al. from the Centers for Disease Control and Prevention in collaboration with the Global Program on AIDS and the World Health Organization. Both the manual and the computer program are in the public domain. Care must be taken when anthropometric measurements are taken to ensure that standardized measurement techniques and calibrated equipment are always used.[23] Measurements should be taken in triplicate, and the mean value recorded.

### 13.3.4 Assessing Levels of Trace Elements and Absorption Modifiers in the Habitual Diets of Study Participants

Information on dietary intakes and on the dietary factors that modify the absorption of trace elements from foods is essential for isotope studies. Such data are required to ascertain the adequacy of the habitual trace-element intakes of the study group(s), the extent to which these intakes may differ from the dietary regimen of the proposed isotope study, and, finally, to measure accurately and precisely the trace-element intakes of the study participants during the metabolic periods. The latter is especially critical when absorption of the trace element depends on the actual quantity in the diet (e.g., zinc and iron).[3]

Assessment of trace-element intakes of the study participants involves three stages: (1) measuring food intakes; (2) converting the intakes of foods to intakes of trace elements and absorption modifiers; and (3) evaluating the adequacy of the trace-element intakes by comparison with reference nutrient intakes. In some circumstances, some of this preliminary background work may have already been performed and highlghted the need for an isotopic study. The three stages are described below.

#### *13.3.4.1 Assessing Food Intakes*

Food intakes can be assessed using quantitative or qualitative dietary assessment methods, depending on the study objectives. When the objective is to measure the intake of trace elements and, where appropriate, absorption modifiers, a quantitative dietary assessment method should be used. In developing countries, the quantitative method that has been most frequently selected is weighed food records, completed in the households by trained dietary monitors. The weighed-food-record method has been described in detail by Gibson.[24] Alternatively, a modified 24-hour recall, especially designed for assessing intakes of trace elements and antinutrients, can be used.[25,26] The modified interactive 24-hour recall method involves training the respondents before the recall, where possible, and incorporates the use of plates, picture charts, and food models to aid respondents in portion-size estimation and recall. The aim of these modifications is to reduce the systematic and random measurement errors by enhancing recall of foods consumed, reducing the number of memory lapses, and improving the portion-size estimates.

#### *13.3.4.2 Compiling a Local Food Composition Table for Use in a Developing Country*

Once the food intake data have been collected, they can then be converted into nutrient and antinutrient intakes. This is usually done by using a food composition table or nutrient database containing values for major and minor elements as well as data for major known absorption enhancers — animal protein, ascorbic acid — and major absorption inhibitors — calcium, phytic acid, dietary fiber, and, where possible, polyphenols. Proximate nutrients such as total protein, fat, and carbohydrate as well as energy should also be included. Details on how to select foods for chemical analysis and how to compile a local food composition table of trace-element values are given in Gibson and Ferguson.[26] Bunch and Murphy have developed an international dietary assessment system suitable for use in many developing countries.[27] This WorldFood Dietary Assessment System includes values for 53 nutrients (including iron, zinc, copper, and manganese) and associated dietary components (including phytic acid and dietary fiber) for 1800 foods consumed in Egypt, Kenya, Mexico, Senegal, India, and Indonesia, as well as a computer program for calculating energy and nutrient intakes.[27]

#### *13.3.4.3 Assessing Intakes of Trace Elements and Absorption Modifiers in Habitual Diets*

Once a suitable food composition table or nutrient database has been located and/or compiled and the food intake data collected, the next step is to calculate the intake of trace elements and antinutrients using a suitable computer software package, such as the WorldFood Dietary Assessment System. Details of this procedure are given by Gibson and Ferguson.[26] The system also computes intakes of available iron and zinc, based on the algorithm of Murphy et al. [28]

To assess the adequacy of trace element intakes, habitual intakes are compared to recommendations. This is usually done by comparing the calculated intakes with tables of recommended nutrient intakes. When tables of recommended intakes for a specific developing country are not available, the tables compiled by the United Nations Agencies — the World Health Organization (WHO) and the Food and Agricultural Organization (FAO) — should be used.[29,30]

For iron and zinc, intakes in developing countries are generally compared to the physiological requirement estimates set by FAO/WHO and WHO, respectively, taking into account the bioavailability of the iron and zinc in the local habitual diets.[29,30] The criteria for classifying local diets as high, moderate, or low for iron and zinc bioavailability have been set by FAO/WHO and WHO, respectively.[29,30] Two levels of requirement estimates for iron and zinc were set. The basal requirement (for iron and zinc) is the level of intake needed to prevent clinically detectable sign of functional impairment. For zinc, the second level is termed the normative requirement estimate, and is the amount needed to maintain a reserve capacity. For iron, the second level is the level of intake needed to prevent anemia in individuals who have evidence of compromised hematopoiesis due to depletion of body iron, but who have not developed anemia. This level is termed the requirement to prevent anemia.

Several methods are available for comparing the calculated trace-element intakes with the physiological requirement estimates set by FAO/WHO and WHO, and are described in detail by Gibson and Ferguson.[26,29,30] All the methods used to evaluate trace-element intakes provide an estimate of the risk of inadequacy of the intake of a trace element at a population and/or individual level. None of the methods actually identify individuals or populations who have a specific nutrient deficiency. This can only be done if biochemical and clinical assessments are also carried out with the dietary investigation.

Once the information on intakes of major minerals and trace elements and absorption modifiers of the population group has been compiled, it can be used to design appropriate test diets for the isotope study, to interpret the response to the test diets, and to calculate appropriate isotope dosages. When estimating the dose required for stable-isotope studies, the aim is to have the lowest possible dose that will achieve the highest reproducibility and relative accuracy, taking into account the sensitivity and precision of the analytical technique, and the estimated efficiency of absorption of the trace element in the study group. Details on calculating the dose for stable isotope studies are given in Chapter 1.

#### *13.3.4.4 Assessing Nutrient Intakes during the Metabolic Study*

Determining the mineral or trace-element content of the test diets accurately and precisely is important because these values, together with the weighed records of food consumed with the isotope administration, are used to calculate the total intake, total absorption, and net absorption of the mineral of interest. The methods and precautions used for assessing food and nutrient intakes during metabolic studies are the same as those used in developed countries. Nevertheless, some of the critical steps are outlined below.

For all isotope studies, the labelled test meals and whole diets *must* be fed to the study participants under close supervision. When studies are carried out on infants, preweighed feeding bottles and bibs must be used to estimate losses during feeding.[31] Furthermore, even when food composition values for local staple foods are available, it is recommended that the trace-element content of the diets consumed over the metabolic periods by each study participant should be determined by direct chemical analysis. In this way, an accurate assessment of the intake of the mineral(s) or trace element under study from the test meal or whole diet can be obtained. Two methods are commonly used: duplicate-diet composites or aliquot sampling of the individual foods consumed.

For duplicate-diet composites, a duplicate portion of all foods and beverages consumed by each study participant is collected during each consecutive 24-hour period of the metabolic collection, whereas for single test meals, a duplicate of the test meal should be collected for analysis. For aliquot sampling, all foods and beverages consumed by each study participant are weighed, and then an aliquot of *each* food and beverage item is collected. These aliquots are then combined into a composite in a trace-element-free blender and subsequently analyzed for mineral or trace-element content. Care must be taken to ensure that representative homogenous aliquots are taken during the duplicate-diet composite collection or aliquot sampling method, as well as in the subsequent pooling, and final homogenization procedures.

When collecting duplicate-diet composites, care must be taken to avoid adventitious contamination during the preparation and analysis of the samples. Precautions include using a blender coated with Teflon and fitted with Teflon blades; an agate ball mill or agate pestle and mortar for grinding; 18 MΩ deionized water; ultrapure reagents; acid-washed glassware; and only trace-element-free polyethylene materials for sample preparation and analysis. Details of the methods used to analyze the analytical samples for major minerals and/or trace elements are given by Helrich.[32] Flame atomic absorption spectrophotometry (AAS) is the most widely used method for mineral and trace-element analysis in food samples but graphite furnace AAS is also suitable. Multi-element methods for trace-element analysis include instrumental neutron activation analysis (INAA), X-ray fluorescence, and inductively coupled plasma spectroscopy (ICP).

## 13.4 Technical Aspects of Implementing Isotope Studies in Developing Countries

Either radioactive or stable isotopes can be used for studies in developing countries; the choice depends on the nutrient, study group, cost, available laboratory facilities, and ethical considerations. Most isotope studies have measured absorption from single meals rather than from whole diets, with the exception of some recent studies comparing the effect of differing amounts of calcium on iron absorption from the whole diet and studies of zinc homeostasis.[33–36] Generally, the meals or diets are labelled extrinsically with the radioactive or stable isotopes. Several techniques are available for measuring absorption and metabolism of trace elements; some are impractical for field studies in developing countries. The suitability of different techniques and practical considerations regarding their use in developing countries are discussed below.

### 13.4.1 Considerations When Selecting the Isotopic Technique

Several different isotope techniques have been developed to investigate trace-element absorption from single meals or whole diets, and may be used in developing countries. However, conducting such studies in areas remote from the base analytical laboratory site may limit the use of some these methods. Issues such as availability of equipment (refrigerators, freezers, ovens, muffle furnaces), trace-element-free equipment and water, technical assistance, the added cost of shipping samples to the base analytical laboratory, and the existence of cultural sensitivities surrounding the collection of biological and metabolic samples may limit the feasibility of some of these methods.

Metabolic balance collections involving fecal and/or urine samples are often an integral part of most isotope studies designed to investigate trace-element absorption from whole diets. Typically, those studies measuring absorption from single meals have used radioisotopes and whole-body counting (zinc, iron) and/or radionuclide uptake by erythrocytes (iron).[37,38] Other procedures include the tissue-retention method for iron, whole-body counting methods using radioactive isotopes, plasma-tolerance curves, and plasma deconvolution. The latter methods are typically used for measurement of fractional absorption from single-test meals. If cultural sensitivities about the collection of biological and metabolic samples exist, methods requiring only a limited amount of these samples are preferable.

#### *13.4.1.1 Fecal Monitoring*

Fecal monitoring is used to measure absorption for those trace elements with a reasonably high fractional absorption (e.g., zinc, copper, selenium), and in some cases for iron, and is discussed in detail in Chapter 4.[39] The single-isotope, fecal-monitoring technique, in which the test meal is labelled with the isotope, has been used but does not measure directly any endogenous losses. Traditional balance techniques can be combined with the use of an intravenously administered isotope dose, or a double isotopic tracer method, which uses an oral isotope dose with the test meal together with an intravenous dose of a different isotope of the same element. These methods allow the direct measurement of endogenous source minerals which are excreted in the feces, and thus permit true absorption to be measured.[40] Details of these methods are discussed in Chapter 4.

In some studies, estimates of endogenous losses have been based on previous work. However, in some situations, these estimates may not be appropriate. For example, in populations where large amounts of dietary phytate are consumed, the phytate may interfere with the reabsorption of the trace element (e.g., zinc) present in intestinal fluids, a mechanism that is important for regulating the conservation of endogenous zinc. In the case of zinc homeostasis, factors which influence fractional absorption of zinc may also affect endogenous fecal losses of zinc, making accurate measurements of both desirable. Thus an isotope method which measures endogenous intestinal losses is recommended with fecal monitoring, whenever adequate resources are available.

Fecal markers must be used for fecal monitoring to identify the fecal samples related to the test meal or diet, and, where possible, ensure complete fecal collection of unabsorbed isotope. Of the various types of markers available, only coloring agents (i.e., brilliant blue) have been employed in stable-isotope studies in developing countries to date.[35,39,41] The coloring agents are simple to use, non-hazardous, and their presence can be readily detected in the stool samples. However, they can only be used to define the beginning and end of the fecal collection period and not to monitor the completeness of the fecal collections. Radio-opaque pellets or rare-earth metals can be combined with coloring agents to determine post-hoc whether complete fecal collections have been made.

Fecal monitoring techniques have been used to study iron absorption in Gambia, and zinc absorption and metabolism in Brazil and China.[35,39,41] In the latter study, the dual-isotope fecal monitoring approach was applied successfully in a study of zinc absorption in a village-based study of Chinese women of child-bearing age whose habitual dietary zinc intakes were marginal.[35] Sian and co-workers measured fractional absorption of an extrinsic zinc isotope label ($^{67}Zn$) by measuring cumulative fecal excretion of nonabsorbed $^{67}Zn$ and endogenous fecal zinc by intravenous administration of a second stable isotope label ($^{70}Zn$).[35] Such an approach can pose difficulties, however, when used in field conditions in developing countries. Incomplete fecal collections may occur arising from loss of sample by the subjects because of poor compliance or termination of the metabolic collections before all the unabsorbed isotopes have traversed through the gut.[42] Such incomplete fecal collections will lead to an overestimate of fractional absorption. As well, this double-isotope approach may not be practical for measuring absorption in studies in some developing countries because administration of isotopes intravenously may be perceived as too invasive and culturally inappropriate, particularly if isotope studies are to be conducted in a community setting. In such cases, the use of single-isotope fecal monitoring may be the most feasible approach.

#### 13.4.1.2 Urinary Monitoring

A recent development has been the measurement of fractional absorption from urine instead of fecal samples. This approach was first used to measure true fractional absorption of calcium, and adapted by Friel and co-workers for zinc and used recently to assess zinc absorption in women during pregnancy and lactation.[43–45] In this method, fractional absorption is based on the ratio of oral to intravenous tracers in the urine, on the assumption that the fractional rates of urinary excretion of the two tracers are identical. To determine the tracer ratio, a 24-hour urine sample collected on the third day after isotope administration, or a spot-urine sample collected on the morning of the fourth day, has been used. The time period over which urine samples must be collected depends on both the element of interest, the objectives of the study, and the selection of the isotopic method.

Use of urinary monitoring techniques can greatly reduce the amount of sample collected, compared to the fecal monitoring technique, and thus reduce the inconvenience associated with the storage and transport of samples. The shorter metabolic period and less rigorous collection protocol may also improve subject compliance.

#### 13.4.1.3 Tissue Retention

To date, tissue retention is only used to measure bioavailability of iron; difficulties with access to the major tissue pools of other minerals preclude the use of tissue retention for measuring absorption of other trace elements. The method is based on the incorporation of radioactive or stable isotopes of iron into red blood cell hemoglobin as described in Chapter 6. It has the advantage of requiring relatively small volumes of sample, which make storage and shipping more convenient. Two different radioactive isotopes of iron ($^{59}Fe$ and $^{55}Fe$) have frequently been used in adults to compare iron absorption between different test meals served on consecutive days in the same adult subjects.[38] Such a design avoids any confounding effect of discrepancies in the initial iron status between the subjects, as discussed in Section 13.3.2, an important advantage in studies in developing countries where, typically, there is a high prevalence of iron deficiency. For radioactive isotope studies using $^{59}Fe$, retention can also be measured by whole-body counting of $^{59}Fe$. However, this method is not practical for field studies in developing countries.

#### 13.4.1.4 Plasma Tolerance Curves and Plasma Deconvolution

Fractional absorption can also be measured using plasma tolerance curves and plasma deconvolution (see Chapter 4). Plasma deconvolution has been used in a clinical study to measure iron absorption in pregnant women.[46] Both methods involve the administration of both oral and intravenous isotopes, and require the insertion of a canula by a clinician so that blood samples can be taken at regular intervals for up to six to eight hours after administration of the isotopically labelled test meal. The intravenous isotope is used as a reference dose and assumes 100% absorption. The concentration of the isotopic labels in the plasma is plotted against time from which the area under the curve is calculated and used to determine fractional absorption of the oral dose. Although these methods are impractical for use in field-based community studies, they may be appropriate for use in clinical settings because of the relatively short subject confinement period, and small volume of sample collected.

### 13.4.2 Collecting, Preparing, and Processing the Metabolic Samples for Analysis of Native Trace Elements and Isotopic Enrichment

It is likely that analysis of native elements and isotopic enrichments in metabolic samples will be carried out in a base analytical laboratory remote from the site of the study. Depending on the cultural beliefs surrounding the collection and use of biological samples, sample collection procedures may require special attention. Depending on the resources available at the study site, some preliminary sample processing may be carried out to reduce the volume and weight of the samples prior to shipment.

Appropriate precautions should be used when handling all types of biological samples because of the high risk of infection. Laboratory equipment, disposal facilities, and working procedures must all be appropriate for working with biologically hazardous materials. Specific considerations for the collection and handling of biological samples in developing countries are discussed below.

#### *13.4.2.1 Fecal Samples*

When the isotope techniques used are based on balance techniques, care must be taken to ensure compliance and complete collection of fecal samples throughout the metabolic period. All stool samples should be collected individually into pre-weighed, 500 to 750 ml, opaque polyethylene containers with a wide opening, or in trace-element free plastic liners. For studies with children, use of a toddler's toilet chair with a trace-element-free plastic bag covering the bowl is recommended. Each stool sample is collected separately, and the collection time, date, and weight are recorded before the sample is frozen.

To calculate the total amount of the trace-element isotopic label excreted in the feces, both the isotopic enrichment and the total elemental content of the native trace element must be measured. This can be done in two ways, depending on the isotope technique used. The isotopic ratios and native trace element can be measured in a homogeneous aliquot from each individual fecal sample collected from each participant. Alternatively, the stool and urine samples collected from each participant over the entire metabolic period can be *pooled*, and a representative subsample withdrawn from each pool for processing and analysis. Note that it is not necessary to take precautions to control for adventitious sources of contamination during the actual collection, pooling, and homogenization of the fecal and urine samples. However, once a subsample of feces or urine is withdrawn for the analysis of isotope enrichment and native trace-element content, then adventitious sources of contamination must be avoided at all stages of the sample preparation and analysis.

When facilities are available within the country, some preliminary sample preparation may be possible. If freeze-dryers are available, all individual frozen stool samples for each participant can be lyophilized individually, directly in the collection container, and then re-weighed. If an appropriate electric oven is available, pooled or individual samples may be dried to constant weight. All powdered stool samples for analysis must be kept in a dissector in a cool, dark environment during storage and shipping. Care must be taken to ensure that samples are re-homogenized before analysis as the sample can segregate during storage or transportation. If it is not possible to pre-process stool samples, then samples should be frozen individually immediately after collection and shipped under dry ice to the base analytical laboratory for sample preparation and analysis.

#### *13.4.2.2 Urine Samples*

When collecting 24-hour urine samples for isotope studies in older children and adults, pre-weighed, wide-neck, polyethylene containers (4-L), with their empty weight, without lid, recorded on the container label, and polyethylene funnels can be used. Acid washed containers are not necessary. Each 24-hour urine sample must be weighed after collection, the weight recorded, and then the sample should be refrigerated immediately. When spot-urine samples, rather than 24-hour urine collections, are required, 1-L polyethylene collection bottles can be used for children and adults. For infants, urine collection bags can be used.

If trace-element-free equipment and an appropriate environment are available, preliminary processing of urine samples can be done prior to shipping. At the end of the pre-defined collection period, each urine sample is thoroughly mixed, the total weight of the urine collection noted, and ten percent by weight withdrawn from each urine sample collection from each study subject and combined into a single pooled urine sample for each subject in a polyethylene bottle. If it is not desirable to pre-process urine samples on site, complete urine samples may be shipped. Care should be taken to use secure containers to avoid spillage during transport.

#### *13.4.2.3 Blood Samples*

For some trace elements (e.g., zinc), blood collection must be taken under carefully controlled, standardized conditions, and refrigerated as soon as possible after collection. For example, the length of time prior to separation of serum/plasma is known to affect zinc concentrations. Changes cannot be detected during the first hour, but longer intervals prior to separation are associated with progressively increasing serum and plasma zinc concentrations.[47] Trace-element-free evacuated containers should always be used for blood samples collected for analysis of native trace elements. When plasma is collected for zinc analysis, use of inappropriate anticoagulants, and hemolysis of the red blood cells must be avoided. After centrifuging, serum or plasma must be separated using trace-element-free polyethylene transfer pipettes, and aliquots stored in trace-element-free polyethylene vials with tight caps. For isotopic analysis, samples can be stored at –10°C; for analysis of native trace elements, samples should be stored at –20°C. Thus it is necessary to have adequate equipment to separate serum/plasma on-site or in nearby facilities, prior to shipping blood samples. Blood samples may be sent under dry ice when the shipping time is relatively short (i.e., less than 8 days) and a specially insulated container is used, or in a liquid-nitrogen-chilled canister to ensure they remain frozen during transport for longer shipping times.

## 13.5 Conclusion

Use of isotopic techniques in developing countries will facilitate a greater understanding of factors associated with the etiology of micronutrient deficiencies and the development of appropriate intervention strategies for their prevention. Isotopic methods are essential for identifying dietary factors that modify the bioavailability of "at risk" nutrients, and for establishing whether any intestinal adaptation occurs with habitual exposure to plant-based diets. Further, the methods can be employed to compare the relative efficiency of absorption of different doses and forms of micronutrient supplements and fortificants, so that optimal doses and forms can be selected for national supplementation and fortification programs to ensure optimal impact. When carrying out isotope studies in developing countries, special consideration should always be given to accessibility of appropriate facilities, equipment for storage and processing, and technical assistance. As well, the expected level of compliance in relation to the cultural setting and age of the subjects must also be taken into account.

To date, very few isotope studies have been performed in developing countries. This is unfortunate because confounding factors such as habitual dietary intakes and trace-element status, the health and physiological status of the individual, presence of adaptive mechanisms controlling mineral homeostatis, and interactions with components in the total diet may impact the results, and hence limit the applicability of studies that have been performed in developed countries. However, in the future, with the development of more sensitive instruments and newer, less invasive techniques, and fueled by the urgency to alleviate micronutrient malnutrition in developing countries, studies employing isotope techniques will be increasingly used.

## References

1. United Nations Administrative Committee on Coordination Sub-Committee on Nutrition (ACC/SCN). Third Report of the World Nutrition Situation. World Health Organization, Geneva, 1997.

2. Sandström, B., Davidsson, L., Eriksson, R., and Alpsten, M., Effect of long-term trace element supplementation on blood trace element levels and absorption of ($^{75}$Se), ($^{54}$Mn), and ($^{65}$Zn), *Journal of Trace Elements and Electrolytes in Health and Disease,* 4, 65, 1990.
3. Sandström, B., Dose dependence of zinc and manganese absorption in man, *Proceedings of the Nutrition Society,* 51, 211, 1992
4. Weigand, E., Kirchgessner, M., and Helbig, U., True absorption and endogenous fecal excretion of manganese in relation to its dietary supply in growing rats, *Biological Trace Element Research,* 10, 265, 1986.
5. Sandström, B. et al., Oral iron, dietary ligands and zinc absorption, *Journal of Nutrition,* 115, 411, 1985.
6. Davidsson, L. et al., Zinc absorption in adult humans: the effect of iron fortification, *British Journal of Nutrition,* 74, 417, 1995.
7. Rossander-Hultén, L. et al., Competitive inhibition of iron absorption by manganese and zinc in humans, *American Journal of Clinical Nutrition,* 54, 152, 1991.
8. FitzGerald, S., Fortification Rapid Assessment Guidelines Tool (FRAT). Ottawa, Path Canada, 1997.
9. MacPhail, A.P. et al., Factors affecting the absorption of iron from Fe EDTA, *British Journal of Nutrition,* 45, 215, 1981.
10. Davidsson, L., Kastenmayer, P., and Hurrell, R. F., Sodium iron EDTA (NaFe (lll) EDTA) as a food fortificant: the effect on the absorption of zinc and calcium in women, *American Journal of Clinical Nutrition,* 60, 231, 1994.
11. Hurrell, R. F., Preventing iron deficiency through food fortification, *Nutrition Reviews,* 55, 210, 1997.
12. Lowe, J.A., Wiseman, J., and Cole, D.J.A., Absorption and retention of zinc when administered as an amino-acid chelate in the dog, *Journal of Nutrition,* 124, 2572S, 1994.
13. Gibson, R.S. and Ferguson, E.L., Nutrition intervention strategies to combat zinc deficiency in developing countries, *Nutrition Research Reviews,* 10, 1, 1998.
14. Engelmann, M.D.M. et al., The influence of meat on nonheme iron absorption in infants, *Pediatric Research,* 43, 768, 1998.
15. Bendich, A. and Cohen, M., Ascorbic acid safety: analysis of factors affecting iron absorption, *Toxicology Letters,* 51, 189, 1990.
16. Charlton, R.W., The effects of organic acids, phytates, and polyphenols on the absorption of iron from vegetables, *British Journal of Nutrition,* 49, 331, 1983.
17. Scott, M.L. and Zeigler, T.R., Evidence for natural chelates which aid in the utilization of zinc by chicks, *Journal of Agricultural Food Chemistry,* 11, 123, 1963.
18. Sandberg, A.-S., Rossander-Hultén, L., and Turk, M., Dietary *Aspergillus niger* phytase increases iron absorption in humans, *Journal of Nutrition,* 126, 476, 1996.
19. Mendoza, C. et al., Effect of genetically modified, low-phytic acid maize on absorption of iron from tortillas, *American Journal of Clinical Nutrition,* 68, 1123, 1998.
20. Cook, J.D., et al., The influence of different cereal grains on iron absorption from infant cereal foods, *American Journal of Clinical Nutrition,* 65, 964, 1997.
21. Gibson, R. S., Determining nutritional status, in *Essentials of Human Nutrition,* Mann, J.I. and Truswell, S., Eds., Oxford University Press, 1998, 427.
22. Dean, A.G. et al., Epi Info, Version 6: a word processing, data base, and statistics program for epidemiology on microcomputers, Centers for Disease Control and Prevention, Atlanta, GA, 1994.

23. Lohman, T. G., Roche, A. F., and Martorell, R. (Eds.), *Anthropometric Standardization Manual*, Human Kinetic Books, Champagne, Il, 1988.
24. Gibson, R. S., *Principles of Nutritional Assessment*, Oxford University Press, New York, 1990.
25. Ferguson, E.L. et al., An interactive 24-hr recall technique for assessing the adequacy of trace mineral intakes of rural Malawian women: its advantages and limitations, *European Journal of Clinical Nutrition*, 49, 565, 1995.
26. Gibson, R.S. and Ferguson, E.L., *An Interactive 24-hour Recall for Assessing the Adequacy of Iron and Zinc Intakes in Developing Countries*, International Life Sciences Institute Press, Washington, D.C., 1999.
27. Bunch, S. and Murphy, S.P., *User's Guide to the Operation of the WorldFood Dietary Assessment Program*, Berkeley, California, Office of Technology Licensing, University of California, 1994.
28. Murphy, S.P., Beaton, G.H., and Calloway, D.H., Estimated mineral intakes of toddlers: predicted prevalence of inadequacy in village populations in Egypt, Kenya and Mexico, *American Journal of Clinical Nutrition*, 56, 565, 1992.
29. WHO (World Health Organization), Trace elements in human nutrition and health, Geneva: World Health Organization, 1996.
30. FAO/WHO (Food and Agricultural Organization/World Health Organization), Requirements of vitamin A, iron, folate and vitamin B-12, FAO, Rome, 1988.
31. Kastenmayer, P. et al., A double stable isotope technique for measuring iron absorption in infants, *British Journal of Nutrition*, 71, 411, 1994.
32. Helrich, K., Ed., *Official Methods of Analysis of the AOAC*, 15th ed., Association of Official Analytical Chemists, Arlington, Virginia, 1990.
33. Gleerup, A., Rossander-Hultén, L., Gramatkovski, E., and Hallberg, L., Iron absorption from the whole diet: comparison of the effect of two different distributions of daily calcium intake, *American Journal of Clinical Nutrition*, 61, 97, 1995.
34. Rossander-Hultén, L. et al., Iron absorption from the whole diet. Relation to meal composition, iron requirements and iron stores, *European Journal of Clinical Nutrition*, 49, 794, 1995.
35. Sian, L. et al., Zinc absorption and intestinal losses of endogenous zinc in young Chinese women with marginal zinc intakes, *American Journal of Clinical Nutrition*, 63, 348, 1996.
36. Hunt, J.R., Matthys, L.A., and Johnson, L.K., Zinc absorption, mineral balance, and blood lipids in women consuming controlled lactoovovegetarian and omnivorous diets for 8 weeks, *American Journal of Clinical Nutrition*, 67, 421, 1998.
37. Anand, B.A., Callender, S.T., and Warner, G.T., Absorption of inorganic and haemoglobin iron in coeliac disease, *British Journal of Haematology*, 37, 409, 1977.
38. Cook, J.D. et al., Food iron absorption measured by an extrinsic tag, *Journal of Clinical Investigation*, 51, 805, 1972.
39. Fairweather-Tait, S.J., Minski M.J., and Singh, J., Nonradioactive method for measuring iron absorption from a Gambian meal, *American Journal of Clinical Nutrition*, 46, 844, 1987.
40. Fairweather-Tait, S. J., Wharf, S. G., and Fox, T. E., Zinc absorption in infants fed iron-fortified weaning food, *American Journal of Clinical Nutrition*, 62, 785, 1992.
41. Jackson, M.J. et al., Stable isotope metabolic studies of zinc nutrition in slum-dwelling lactating women in the Amazon valley, *British Journal of Nutrition*, 59, 193, 1988.

42. King, J.C. et al., The double isotope tracer method is a reliable measure of fractional zinc absorption, *European Journal of Clinical Nutrition*, 51, 787, 1997.
43. Yergey, A.L., Vieira, S.A., and Covell, D.G., Direct measurement of dietary fractional absorption using calcium isotopic tracers, *Biomedical and Environmental Mass Spectrometry*, 14, 603, 1987.
44. Friel, J.K. et al., The analysis of stable isotopes in urine to determine the fractional absorption of zinc, *American Journal of Clinical Nutrition*, 55, 473, 1992.
45. Fung, E.B. et al., Zinc absorption in women during pregnancy and lactation: a longitudinal study, *American Journal of Clinical Nutrition*, 66, 80, 1997.
46. Whittaker, P.G., Lind, T., and Williams, J.G., Iron absorption during normal human pregnancy: a study using stable isotopes, *British Journal of Nutrition*, 65, 457, 1991.
47. English, J.L. et al., Evaluation of some factors that may affect plasma or serum zinc concentrations, in *Trace Elements in Man and Animals* 6, Hurley, L.S. et al., Eds., Plenum Press, New York-London, 1988, 459.

# Index


## A

A priori identifiability, 51
AAS, *see* Atomic absorption spectrometry
Absorption, *see also* True absorption
  chromium, 160
  copper, 38, 66–68
    –zinc, 180–181
  definition, 61–62
  indirect measurements, 35–39
  iron, 62–66, 94
  manganese, 152, 155–156
  selenium, 72–76, 131
  stable isotopes and nutritional research, 12, 14
  trace-element analysis methods, 60–61
  whole-body counting, 30–31
  zinc, 68–72, 189
Accelerator mass spectrometry (AMS), 39
Accuracy, detection methods, 55, 121
Active transport, 131
Administration, *see also* Intravenous administration; Oral administration
  compartmental modelling, 52
  selenium, 141
  stable isotopes, 12–13, 14
Age, 172, *see also* Elderly; Infants
Air exchange system, 33
Albumin, 67
Alkaline phosphatase, 110
AMS, *see* Accelerator mass spectrometry
Analysis of covariance (ANOVA), 201
Analytical sites, 4
ANOVA, *see* Analysis of covariance
Anthropogenic radioisotopes, 27
Apparent absorption, 61, *see also* Absorption
Arch-geometry, 28, 31
Area under the curve (AUC), 65, 71, 178
Atomic absorption spectrophotometry (AAS), 122, 206
AUC, *see* Area under the curve
Availability, 27, 97, 103–104, *see also* Bioavailability

## B

Background radiation, 28, 33
Between-group design, 200, 201, *see also* Developing countries, trace-element studies
Binding protein, 137
Bioavailability
  iron, 168, 170–171, 173, 176–177
  manganese, 153
  selenium, 73, 131–132
  trace elements in developing countries, 194, 196, 198–199, 204
  zinc, 70
Biochemistry, 152, 160, 161
Biological marker, 160
Biological samples, 16, 99–100, *see also* Blood; Fecal monitoring; Urinary monitoring; Urine
Biological turnover rate, 29–30, 34
Biosynthesis, 132
Blood
  copper, 87
  iron, 100
  manganese, 154
  samples and developing countries, 211–212
  selenium, 138, 143
  zinc, 15, 16
Body imaging techniques, 34–35
Body pools, 140–143
Breast feeding
  chromium, 183
  iron, 174–176, 177
  zinc, 179
    copper, 181

## C

Caco-2 cell system, 65, 72
Calcium, 31, 36, 37
Cancer, 134
Carbon, 39
Cereal, 176, 179, 197
Ceruloplasmin, 118
Chemistry, 160, 161
Chromium, 38, 159–161, 182–183

Coefficient of variation, 55
Collimator, 34
Commercial availability, isotopes, 7–8
Community leaders, 200, *see also* Developing countries, trace-element studies
Compartmental analysis, 140, 142
Compartmental modelling, 83–85, 158, 159, *see also* Tracer-to-tracee ratio, compartmental modelling
Complementary foods, 176–177, 179
Compliance, 70
Concrete chambers, 33
CONSAM, 87, *see also* Software
Contamination, 28, 210
Copper
  absorption analysis, 66–68
  biological turnover rate and whole-body counting, 29
  elderly studies, 189–190
  metabolism, 118–123
    fecal monitoring, 38
  stable-isotope research, 6
  whole-body kinetic studies, 86–88
  –zinc absorption in infants, 180–181
Cost, 8, 10, 96
Criteria, stable-isotope studies, 188, 200, 201
Cultured cells, 153–154
Cytochrome c oxidase, 118

## D

Daughter nuclides, 25–26, *see also* Radioisotopes
Decay mode, 25, 26
Deconvolution, 35, 209
Deficiency
  copper, 118
  developing countries, 194, 195, 198–199
  manganese, 152
  selenium, 133–134
  zinc, 109–110, 112
Detoxification, 132
Deuterium, 2
Developing countries, trace-element studies
  dietary strategies, 198–199
  fortification, 197–198
  isotope supplements, 195–197
  practical aspects, 199–206
  technical aspects, 206–212
Dietary copper, 87, *see also* Copper
Dietary fiber, 179
Dietary strategies, 198–199, *see also* Developing countries, trace-element studies
Dietary zinc, 110, *see also* Zinc
Diets
  duplicate, 209
  habitual and trace elements in developing countries
  absorption modifiers, 203–206
  supplement use, 196
  plant-based and zinc/copper absorption, 181
Disease, 133–134, 188
Disposal costs, 4
Distribution, 111, 152, 154-155
Dosage
  kinetic studies of whole-body trace elements, 82
  selenium isotope measurement, 140
  stable isotopes, 12–13, 14
    infants/pregnant/lactating women, 168
Double-isotope technique, *see* Dual isotope studies
Dual isotope studies
  developing countries, 201, 208
  dosage and administration, 12
  iron, 63, 65, 102, 104
  tracer-to-tracee ratio, 53
  zinc, 16, 69
Duplicate diet composites, *see* Diets

## E

EIMS, *see* Electron ionization mass spectrometry
Elderly, 188–190
Electron capture, 25
Electron ionization mass spectrometry (EIMS), 120–121
Endogenous constant steady state, 54
Energy, 25, 26
Enrichment
  detection, 4, 6, 8, 12–13
  iron, 96, 98, 103
  tracer-to-tracee ratio, 48
Enzymic techniques, 138
Equipment, 31–34, 38–39
Errors, 28, 104
Erythrocytes, *see also* Red blood cells
  glutathione peroxidase assays, 135, 138, 139
  iron incorporation
  infants, 171, 173–177
  pregnant women, 177–178
Erythropoiesis, 104
Essentiality
  chromium, 160
  manganese, 153–159
Ethics, 188–189
Exchangeable pool, 88–89, 141–142, 190, *see also* Selenium
Exchangeable zinc pool (EZP), 113
Excretion, 30, 155, 181

Extrinsic labelling, 31, 175, *see also* Intrinsic labeling
EZP, *see* Exchangeable zinc pool

## F

FAB-MS, *see* Fast atom bombardment mass spectrometry
FAO, *see* Food and Agricultural Organization
Fast atom bombardment mass spectrometry (FAB-MS), 3, 11, 14, 100
Fat, 39
Feasibility issues, 95–99
Fecal markers, 208
Fecal monitoring
- copper 66–67
- developing countries, 207–208
- collection, preparation, processing, 210–211
- indirect measurements of absorption/metabolism, 38
- infants/pregnant/lactating women, 169–170
- iron, 64, 97–98, 175
- manganese, 155
- selenium, 74, 182
- zinc, 16, 69–70, 179

Feeding patterns, 168–169
Fission, 27
Food and Agricultural Organization (FAO), 204
Food intake, 203–204, *see also* Developing countries, trace-element studies
Formula feeding, *see also* Breast feeding
- iron, 172, 174–176, 177
- selenium, 181–182
- zinc, 179
- zinc/copper, 181

Fortification, foods, 197–198, *see also* Developing countries, trace-element studies
Fraction correction, 18

## G

Gallium, 17
Gas chromatography–mass spectrometry (GC–MS), 8, 9, 120–121, 139, 140
Gastrointestinal tract, 136
GC–MS, *see* Gas chromatography–mass spectrometry
Geiger-Müller counter, 31
Genetic defects, 118
Germanium detectors, 32, 38
Glutathione peroxidases (GSHpx), 135–136
GSHpx, *see* Glutathione peroxidases

## H

Half-life, 3, 24, 26
Health, 201–203, *see also* Developing countries, trace-element studies
Heart, 135
Hematocrit, 103
Hemoglobin, iron content, *see also* Erythrocytes; Red blood cells
- absorption analysis, 63–64
- calculations for infants, 171
- feasibility issues using stable isotopes, 95, 98
- mucosal absorption, 102–103

Hereditary hemochromatosis (HHC), 103
HHC, *see* Hereditary hemochromatosis
History, 1–3, 188
Homeostasis
- copper, 87
- manganese, 153
- zinc, 15, 86, 180

Human ethics committee, 199–200
Human studies, 2, 8, 161, 154–155

## I

ICP-MS, *see* Inductively coupled plasma mass spectrometry
*In vitro* methods, 65, 72
Indistinguishability principle, 45, 51
Inductively coupled plasma mass spectrometry (ICP-MS)
- copper, 121
- iron 63–64, 100, 101
- kinetic studies, 81
- mass spectrometry techniques, 2, 10–11
- precision and nutritional research, 13
- selenium, 73–74, 139–140
- zinc, 16–17

Infants
- chromium, 183
- iron, 63, 102–103, 104, 171–172, 173–177
- trace-element metabolism studies, 3
- zinc, 179
  - –copper, 180–181

Instrumentation, stable-isotope research, 8–11
Intake
- iron, 94
- selenium, 130–131, 143–145

Interference, 67, 140

Intervention strategies, 194, 195–197, *see also* Developing countries, trace-element studies
Intravenous administration, *see also* Oral administration
  iron status, 17, 18, 65, 96, 171–172, 173
  nutritional research, 12, 14
  practical strategies and stable-isotope studies, 15
  zinc status, 112, 179–180
Intrinsic labeling, *see also* Extrinsic labeling
  nutritional research, 14
  selenium, 75, 132, 143, 145
  trace-element absorption, 60
  whole-body counting, 31
Iodothyronine deiodinases, 136
Ionization chamber, 31
Iron
  absorption analysis, 62–66
  developing countries
    dietary strategies, 198
    intake assessment, 205
    supplements, 195–196
  erythrocyte incorporation and absorption
    pregnant women, 177–178
    premature infants, 173–174
    term infants, 174–177
  heme/non-heme absorption, 62
  indirect measurements of absorption, 35
  inductively coupled plasma mass spectrometry in research, 10
  –manganese interactions, 153, 158–159
  metabolism and stable-isotope methods
    analytical methods, 99–102
    applications, 102–104
    feasibility issues, 95–99
    protocol design, 94–95
  stable-isotope research, 2–3
    practical strategies, 17–18
  trace-element study methodology, 170–173
  whole-body counting, 29, 30

## K

Kidney, 136
Kinetic modelling, 140–143
Kinetics studies, 123
  compartmental modelling, 83–85
  considerations in study design, 82–83
  copper, 86–88
  selenium, 88–89
  zinc, 85–86

## L

Laboratory animals, 153–154
Laboratory tests, 202
Lactating women, 180, 182–183
Lactoferrin, 174, 175
Limit of detection (LOD), 10, 13, 32, 63, 71
Limit of quantification (LOQ), 71
Local food composition table, 204, *see also* Developing countries, trace-element studies
LOD, *see* Limit of detection
LOQ, *see* Limit of quantification
Low phytic acid maize (LPM), 198–199
LPM, *see* Low phytic acid maize

## M

Manganese
  biological turnover rate and whole-body counting, 29
  isotope studies
    biochemistry, 152
    essentiality, 153–159
  radiation dose and absorption studies, 27
  supplement use in developing countries, 196
Mass balance equation, 44–45, 50
Mass spectrometry (MS)
  advances in stable-isotope research, 2–3
  iron metabolism stable-isotope studies, 100–102
  tracer-to-tracee ratio calculation, 47–50
Mathematical modelling, 156, 157–158
MCS, *see* Multichannel scanning
Measuring time, 31
Menkes' disease, 118
Metabolic beds, 170
Metabolism, *see also* Trace-element metabolism, kinetic studies of whole-body
  compartmental modelling, 43
  indirect measurements, 35–39
  nutrient intake assessments in developing countries, 205–206
  radiation dose, 27
  selenium, 132–133
  mass spectrometry methods, 2
  trace-element and stable isotopes, 3–11
  whole-body counting applications, 29–30
  zinc, 15, 180
Metallothionein, 110–111
Methylation, 132

Mitochondrial capsules, 137
Modified 24-hour recall, 204
Molybdenum, 161–162
Monte Carlo methods, 55
MS, *see* Mass spectrometry
Mucosa, 102–103
Multichannel scanning (MCS), 34–35
Multi-compartmental studies, 85–86
Multiple labelling, 121–122, *see also* Extrinsic labeling; Intrinsic labeling
Multi-pool tracer, 50–53
Myocarditis, 133–134

## N

NAA, *see* Neutron activation analysis
NaI detectors, 31, 32, 33, 38
National Center for Health Statistics, 203
NCHS, *see* National Health Statisics
Neutron activation analysis (NAA)
  iron metabolism, 99–100
  mineral stable-isotope research, 2, 9, 14–15
  copper satus and metabolism, 120
Nitrogen, 2
Non-heme iron, *see* Iron
Normalization of daa, 62–63
Nutrition
  chromium, 160–161
  development countries, 201–203
  dosage ranges of stable isotopes, 12–13
  stable-isotope methodology, 2

## O

Old steel chambers, 28, 32
One-compartment model, 112
Oral administration, *see also* Intravenous administration
  iron absorption, 17, 65, 96, 97
  nutritional research with stable isotopes, 12
  zinc, 15
Orogastic feeding tube, 173
Oxidative hemolysis, 135

## P

PEG marker, 71–72
PHA, *see* Pulse Height Analysis
Phospholipid hydroperoxide, 135
Photons, 25
Physical examinations, elderly, 188
Phytase, 198
Phytate, 207–208
Plasma, 33, 98–99, 136, 209
Plasma appearance/disappearance techniques
  copper analysis, 67, 87
  iron analysis, 65
  selenium analysis, 74–75, 88
  zinc analysis, 70–71
Plastic detectors, 33
Platelets, 138
Polypharmacy, 188
Pooled samples, 210
Potassium, 33
Precision
  instrumentation and nutritional research, 12
  neutron activation analysis, 100
  thermal ionization mass spectrometry, 9, 121
  TTRM software and compartmental modelling, 55
Pregnant women
  iron absorption, 177–178
  metabolism study and use of stable isotopes, 3
  selenium absorption, 143, 182
  zinc/copper absorption, 181
Premature infants, 179, *see also* Infants
Protected fortificants, *see* Fortification
Pulse Height Analysis (PHA), 34
Purity, stable isotopes, 3

## R

Radiation dose, 25–27
Radioimmunoassays, 138
Radioisotopes
  body imaging techniques, 34–35
  characteristics, 24–27
  copper status, 119–120
  indirect measurements, 35–39
  iron bioavailability, 196
  manganese essentiality, 153–159
  molybdenum studies, 161–162
  selenium absorption, 74
  stable isotopes comparison, 60
  whole-body counting techniques, 27–34
  zinc status, 111–112
Radionuclides, *see* Radioisotopes
Radium, 31
Rare-earth elements, 64, 67, 74
RBC, *see* Red blood cell
RDA, *see* Recommended daily allowance
Recommended daily allowance (RDA), 152
Red blood cells (RBCs), 17, 63, 95–96, 98, *see also* Erythrocytes
Reexcretion studies, 69, 170
Reference-dose absorption, 62–63

Relative standard deviation (RSD), 139, 140
Retention, 35, 154–155, 209
Risks, 205, *see also* Developing countries
RSD, *see* Relative standard deviation

## S

SAAM, 142, *see also* Software
SAAM/CONSAM, 85–86, 88, 112, *see also* Software
SAAMII, 71, *see also* Software
Sample collection, 15, 16
Sample preparation, 4
Scanning-bed geometry, 32, 33
Se-EMP, *see* Selenite-exchangeable metabolic pool
Selenite-exchangeable metabolic pool (Se-EMP), 88–89, *see also* Selenium
Selenium
  absorption analysis, 72–76
  biological turnover rate, 29
  extrinsic labeling and nutritional research, 14
  stable-isotope studies
    assessment of status, 137–145
    dietary and metabolism, 130–133
    elderly, 190
    infants/pregnant/lactating women, 181–182
    role in body, 133–134
    selenoproteins, 134–137
    urinary excretion and indirect measurements, 36
    whole-body kinetic studies, 88–89
Selenoproteins, 134–137
Serum ferritin, 62, 170, *see also* Iron
Shielding, 31
Simulation models, 71
Single stable-isotopes techniques, 69
Single-pool tracer, 44–47, 55
Software
  compartmental modelling, 54–55, 83–84, 85
  selenium exchangeable pool, 142
  zinc status determination 112
Soil, 131
Soy formula, 175–176, *see also* Formula feeding
Spectral interference, 10, *see also* Interference
Stable-isotope studies
  chromium, 161
  copper, 120–123
  cost, access, and limitations of use, 168
  dosage, preparation, and administration, 12–14
  elderly, 188–190
  historical background, 1–3
  iron, 93–95, 171–172, 177–178
  molybdenum studies, 162
  practical strategies for conducting studies, 14–18
  radioisotopes comparison, 60
  selenium, 139–140, 143–145
  suppliers, 18–19
  trace-element metabolism, 3–11
  zinc, 180
    –copper, 181
Status assays
  selenium, 137–145, 190
  copper, 189–190
Stores, iron, 94–96
Stress, 110
Study design, 82–83, 200–201
Study protocols, compliance, 168
Superoxide dismutase, 118
Supplements, 144, 195–197
Suppliers, 4, 18–19
Support, 199–200, *see also* Developing countries, trace-element studies

## T

Technological development, 31–34, 38–39
Test meals, 168–169
Thermal ionization mass spectrometry (TIMS)
  advances in stable-isotope research, 2, 9–10
  copper status, 121
  iron studies, 17, 100
  precision and nutritional research, 13
Thioredoxin reductase, 136
Thyroxin, 136
Time series designs, 200, *see also* Developing countries, trace-element studies
TIMS, *see* Thermal ionization mass spectrometry
Tissue retention, *see* Retention
Tissues, 38
Total element analysis, 9
Toxicity, 24, 130, 152
Trace elements
  absorption analysis
    copper, 66–68
    general description, 60–62
    iron, 62–66
    selenium, 72–76
    zinc, 68–72
  developing countries intake assessment, 204–205
  infants/pregnant/lactating women
    background and overview, 167–170
    chromium, 182–183
    iron, 170–178

selenium, 181–182
zinc and copper, 180–181
zinc, 178–180
kinetic studies of whole-body
compartmental modelling, 83–85
considerations in study design, 82–83
copper, 86–88
selenium, 88–89
zinc, 85–86
Tracer-to-tracee ratio (ttr), compartmental modelling
mass spectrometry measurements, 47–50
multiple tracer case, 52–53
multi-pool tracer kinetics, 50–52
single-pool tracer kinetics, 44–47
test of the endogenous-constant, steady-state assumption, 54
TTRM software, 54–55
whole-body kinetic studies, 83
Transcellular transport, 153
Transferrin, 98
Triple-isotope technique, 18, 104
True absorption, *see also* Absorption
definition, 61
manganese, 156–158
selenium, 75
zinc, 69
ttr, *see* Tracer-to-tracee ratio
TTRM software, 54–55
Tumorigenesis, 134
Two-compartment model, 112–113, *see also* Compartment modelling


## U


UNICEF, *see* United Nations Children's Fund
United Nations Children's Fund (UNICEF), 194, 196
Urinary excretion, 36–37, 143, 144
Urinary monitoring
chromium, 182
copper, 118
selenium, 75, 182
trace-element studies in developing countries, 208–209
zinc, 70, 86
Urine, samples
trace-element studies in developing countries, 211, 209
measurement using NaI detectors, 38
zinc studies, 15, 16


## V


Valence states, 160, 161
Very-low-birth weight infants, 179–180, 181, *see also* Infants
Viral disease, 133–134
Vitamin supplements, 177, 178


## W


WBC, *see* Whole-body counts
WD gene, 87
Weighted-food-record method, 203
Wheat, 131
WHO, *see* World Health Organization
Whole-body counts (WBC)
applications, 29–34
characterization, 28
copper, 67–68
iron, 64
manganese, 153, 154, 156
mathematical modelling, 156, 157–158
selenium, 75
zinc, 68–69
Whole-gut lavage technique, 71–72
Wilson's disease, 87, 120, 118
Within-group design, 200
World Health Organization (WHO), 194, 204
WorldFood Dietary Assessment System, 204


## Z


Zinc
absorption analysis, 68–72
indirect measurements, 35
biological turnover rate and whole-body counting, 29
developing countries, 196, 205, 208–209
dual-isotope tracer studies, 12
erythrocyte incorporation of iron in premature infants, 173–174, 178–180
exchangeable pool, 113
fast atom bombardment mass spectrometry, 11
fecal monitoring, 38, 208
homeostasis in elderly, 189
metabolism and whole-body kinetic studies, 85–86
radiation dose, 27
status and isotope use, 109–113
strategies for conducting stable-isotope tracer studies, 14–17
tracer-to-tracee ratio calculation, 47, 48, 49
urinary excretion, 36
whole-body kinetic studies, 83, 84